How Might Life Evolve
on Other Worlds?

Life in the Universe Series

How Might Life Evolve on Other Worlds?

SETI Academy Planet Project

SETI INSTITUTE

1995

TEACHER IDEAS PRESS
A Division of
Libraries Unlimited, Inc.
Englewood, Colorado

TEACHER IDEAS PRESS
A Division of Libraries Unlimited, Inc.
P.O. Box 6633
Englewood, CO 80155-6633
1-800-237-6124

Series Production Editor: Kevin W. Perizzolo
Series Copy Editor: Jason Cook
Series Proofreader: Ann Marie Damian
Series Typesetting and Interior Design: Judy Gay Matthews

Library of Congress Cataloging-in-Publication Data

How might life evolve on other worlds? : SETI academy planet project / SETI Institute.
   xxvi, 223p. 22x28 cm. -- (Life in the universe series)
   Includes bibliographical references and index.
   ISBN 1-56308-325-6
   1. Life on other planets--Study and teaching (Elementary) 2. Life on other planets--Study and teaching (Elementary)--Activity programs. 3. Evolution (Biology)--Study and teaching (Elementary) 4. Evolution (Biology)--Study and teaching (Elementary)--Activity programs. I. SETI Institute. II. Series.
QB54.H68 1995
372.3'5--dc20
                                            94-46537
                                              CIP

Contents

Scope and Sequence
Life in the Universe Curriculum

This scope and sequence is designed to describe the topics presented and the skills practiced in the Life in the Universe series curriculum as they relate to factors in the Drake Equation: $N = R_* \cdot f_p \cdot n_e \cdot f_l \cdot f_i \cdot f_c \cdot L$. In this equation, N is an estimate of the number of detectable civilizations in the Milky Way Galaxy that have developed the ability to communicate over interstellar distances. If the number of civilizations with this ability is sufficiently large, then there is an opportunity for us to find them by "eavesdropping" on their communications. This was the rationale for formulating the Drake Equation, and this is the rationale for the search for extraterrestrial life.

Factors in the Drake Equation	Related Topics
R_* = the number of new stars suitable for the origin and evolution of intelligent life that are formed in the Milky Way Galaxy each year	*Astronomy, Chemistry, Mathematics*
f_p = the fraction of these stars that are formed with planetary systems	*Astronomy, Mathematics, Physics*
n_e = the average number of planets in each such system that can sustain life	*Astronomy, Biology, Chemistry, Ecology, Physics*
f_l = the fraction of such planets on which life actually begins	*Astronomy, Biology, Chemistry, Ecology, Geology, Meteorology*
f_i = the fraction of life-sustaining planets on which life evolves	*Anthropology, Biology, Geology, Meteorology, Paleontology*
f_c = the fraction of systems of intelligent creatures that develop the technological means and the will to communicate over interstellar distances	*Language Arts, Mathematics, Physics, Social Sciences*
L = the average lifetime of such civilizations in a detectable state	*Astronomy, History, Mathematics, Paleontology, Social Sciences*

Life in the Universe Series	Topics	Skills
Grades 3-4 *The Science Detectives*	• Art • Astronomy • Chemistry • Language Arts • Mathematics • Physics	• Attribute Recognition • Cooperative Learning • Mapping • Measurement • Problem Solving • Scientific Process
Grades 5-6 *The Evolution of a Planetary System*	• Art • Astronomy • Biology • Ecology • Geography • Geology • Language Arts • Mathematics • Meteorology • Social Sciences	• Problem Solving • Cooperative Learning • Scientific Process • Mapping • Measurement • Inductive Reasoning • Graphing
Grades 5-6 *How Might Life Evolve on Other Worlds?*	• Art • Biology • Chemistry • Ecology • Language Arts • Mathematics • Paleontology • Social Sciences	• Classification • Inductive Reasoning • Laboratory Technique • Mapping • Microscope Use • Scientific Process • Cooperative Learning
Grades 5-6 *The Rise of Intelligence and Culture*	• Anthropology • Art • Biology • Ecology • Geography • Geology • Language Arts • Mathematics • Social Sciences • Zoology	• Creative Writing • Graphing • Laboratory Technique • Mapping • Problem Solving • Cooperative Learning
Grades 7-8 *Life: Here? There? Elsewhere? The Search for Life on Venus and Mars*	• Art • Astronomy • Biology • Chemistry • Comparative Planetology • Ecology • Engineering • Language Arts • Mathematics • Physics • Zoology	• Cooperative Learning • Design • Graphing • Inductive Reasoning • Laboratory Technique • Microscope Use • Problem Solving • Scientific Process
Grades 8-9 *Project Haystack: The Search for Life in the Galaxy*	• Anthropology • Art • Astronomy • Biology • Chemistry • Ecology • Geometry • Language Arts • Mathematics • Physics • Trigonometry • Zoology	• Cooperative Learning • Design • Graphing • Inductive Reasoning • Laboratory Technique • Microscope Use • Problem Solving • Scientific Process

Foreword

Carl Sagan, Cornell University

The possibility of life on other worlds is one of enormous fascination—and properly so. The fact that it's such a persistent and popular theme in books, television, motion pictures, and computer programs must tell us something. But extraterrestrial life has not yet been found—not in the real world, anyway. Through spacecraft to other planets and large radio telescopes to see if anyone is sending us a message, the human species is just beginning a serious search.

To understand the prospects, you need to understand something about the evolution of stars, the number and distribution of stars, whether other stars have planets, what planetary environments are like and which ones are congenial for life. Also required are an understanding of the chemistry of organic matter—the stuff of life, at least on this world; laboratory simulations of how organic molecules were made in the early history of Earth and on other worlds; and the chemistry of life on Earth and what it can tell us about the origins of life. Include as well the fossil record and the evolutionary process; how humans first evolved; and the events that led to our present technological civilization—without which we'd have no chance at all of understanding and little chance of detecting extraterrestrial life. Every time I make such a list, I'm impressed about how many different sciences are relevant to the search for extraterrestrial life.

All of this implies that extraterrestrial life is an excellent way of teaching science. There's a built-in interest, encouraged by the vast engine of the media, and there's a way to use the subject to approach virtually any scientific topic, especially many of the most fundamental ones. In 1966, the Soviet astrophysicist I. S. Shklovskii and I published a book called *Intelligent Life in the Universe,* which we thought of as an introduction to the subject for a general audience. What surprised me was how many college courses in science found the book useful. Since then, there have been many books on the subject, but none really designed for school curricula.

These course guides on life in the universe fill that need. I wish my children were being taught this curriculum in school. I enthusiastically recommend them.

Preface

Are we alone in the Milky Way Galaxy? Many people think of science fiction stories or tabloid reports about UFO abductions when they hear about the search for intelligent life on other planets. The reality is that many scientists take seriously the possibility of life on other worlds, and some have undertaken the difficult task of finding out if we are the only intelligent beings in our galaxy. Astronomer Frank Drake proposed an equation to estimate the number of civilizations in our galaxy that produce radio waves. We might be able to detect such civilizations with our radio telescopes. The Drake Equation estimates this number using the answers to the following sequence of questions:

1. How many stars are formed in the Milky Way Galaxy each year?

2. What fraction of stars are similar to our Sun?

3. What fraction of stars are formed with a planetary system?

4. What is the average number of planets in such a system?

5. What fraction of planets are like Earth, capable of sustaining life?

6. On what fraction of these planets does life actually begin?

7. On what fraction of life-sustaining planets does life evolve into intelligent civilizations?

8. What fraction of intelligent civilizations develop radio technology?

9. What is the average lifetime of a radio-transmitting civilization?

Scientists pursuing these questions work in many fields, including astronomy, geology, biology, anthropology, and the history of science. Several projects to "listen" for radio signals produced by civilizations on distant planets have been conducted. The most ambitious of these has been undertaken by the research staff at the SETI Institute (Search for Extraterrestrial Intelligence), at first in cooperation with NASA (National Aeronautics and Space Administration) and later using privately donated funds. The SETI team is listening for intelligent signals. The interdisciplinary makeup and highly motivational nature of the search for intelligent life prompted the NSF (National Science Foundation) to support the development of the Life in the Universe Curriculum Project. Designed by curriculum developers working with teachers and NASA and SETI scientists, this program reflects the real-life methods of science: making observations, performing experiments, building models, conducting simulations, changing previous ideas on the basis of new data, and using imagination. It brings into the classroom the excitement of searching for life beyond Earth. This search is a unifying theme that can unleash the imagination of students through integrated lessons in the physical, life, space, and social sciences.

The *SETI Academy Planet Project* consists of three books, each of which is a teacher's guide for grades 5-6. *The Evolution of a Planetary System* examines an important aspect of the search for intelligent life: the evolution of stars and planets. Students visualize how our Sun and its family of nine planets have formed and evolved into the solar system we know today. By applying what they have learned about the evolution of Earth, students imagine how planets might have formed around other stars, how individual planets might have evolved through similar processes, and what such planets might look like today. They explore how Earth has changed over time, how tectonic forces deep inside our planet brought about these changes, and how geographic locations and geologic landforms influence climate. Students use the results of their research to design planetary systems that contain habitable planets, "evolve" individual planets into life-sustaining worlds, and create continental and climate maps of their planets.

How Might Life Evolve on Other Worlds? focuses on the vast expanses of time during which plant and animal life evolved on Earth. Students participate in a series of multidisciplinary activities to analyze the origin and evolution of life on Earth. Students discover that life evolved through interaction with the environment and that some life evolved from simple to complex forms. By applying what they have learned about the evolution of life-forms on Earth, students imagine realistic scenarios for how life might evolve on another planet.

The Rise of Intelligence and Culture emphasizes how intelligence and culture helped humans to form a civilization that now has the technology to detect and communicate with possible extraterrestrial civilizations. Students learn about indications and characteristics of intelligence, about the evolutionary increase in the size of the human brain, about survival needs, and about the stages of human culture. They examine the possibility of sending messages through space and the social issues related to the search for extraterrestrial intelligence. By applying what they have learned, students contemplate how an extraterrestrial civilization might have evolved.

The *SETI Academy Planet Project* provides an exciting, informative, and creative series of activities for elementary students, grades 5-6. In these activities, each student plays the role of a cadet at the SETI Academy, a fictitious institution. Each book of the *SETI Academy Planet Project* is designed to be a complete unit in itself as well as a subunit of a three-unit course. The use of these guides rests with each teacher.

 # Curriculum Development Team

Principal Investigator:	Dr. Jill Tarter, SETI Institute, Mountain View, CA
Project Director:	Dr. David Milne, Evergreen State College, Olympia, WA
Project Evaluator:	Dr. Kathleen A. O'Sullivan, San Francisco State University
Curriculum Development Manager:	Cara Stoneburner, SETI Institute
Editor:	Victoria Johnson, SETI Institute, San Jose State University, CA
Contributing Authors:	Kevin Beals, Lawrence Hall of Science, Berkeley, CA Mary Chafe-Powles, Woodside Elementary School, Concord, CA Lisa Dettloff, Lawrence Hall of Science Alan Hewitt, Lawrence Hall of Science Victoria Johnson, SETI Institute Betty Merritt, Longfellow Intermediate School, Berkeley, CA Dr. David Milne, Evergreen State College Dr. Cary Sneider, Lawrence Hall of Science Cara Stoneburner, SETI Institute Emily Theobald, SETI Institute Lisa Walenceus, Lawrence Hall of Science
AV Consultants:	Jon Lomberg, Honaunau, HI Dr. Seth Shostak, SETI Institute
Research Assistants:	Winslow Burleson, SETI Institute Amy Barr, student at Palo Alto High School, CA Lisa Chen, student at Palo Alto High School, CA Ladan Malek, San Lorenzo High School, CA
Evaluator:	Jennifer Harris, Educational Consultant, Redwood Valley, CA
Artist:	Jon Lomberg, Honaunau, HI Stuart Timmons, Los Angeles
Poster:	Jon Lomberg, Honaunau, HI
Video Image Show:	Anna Domitrovic, Arizona-Sonora Desert Museum, Tucson Jon Lomberg, Honaunau, HI Dr. David Milne, Evergreen State College David Thayer, Arizona-Sonora Desert Museum, Tucson Judith Weik, Arizona-Sonora Desert Museum, Tucson Simon Bell Production Services, Toronto
Advisory Board:	Tom Pierson, SETI Institute Dr. Peter Backus, SETI Institute Edna K. DeVore, SETI Institute Dr. Gilbert Yanow, Jet Propulsion Laboratory, Pasadena, CA

Acknowledgments

Development and publication of the Life in the Universe series was made possible by grants from the National Science Foundation (grant #MDR-9510120), and the National Aeronautics and Space Administration (grant #NCC-2-336). This support does not imply responsibility for statements or views expressed in this publication.

Field Test Teachers

Teacher	School
Georgia D. Albritton	Timberwilde Elementary School, San Antonio, TX
Christine Anderson	Spring Grove School, Hollister, CA
Stephen Bennett	Radnor Middle School, Wayne, PA
Jo Anne Butler	Del Dios Middle School, Escondido, CA
Barbara Cardwell	C. D. Fulkes Middle School, Round Rock, TX
Victor Chang	Waiau Elementary School, Pearl City, HI
Helen Cole	Iditarod Elementary School, Wasilla, AK
Patricia Harvey	Oak Knoll School, Menlo Park, CA
Linda Kirk	Almond Elementary School, Los Altos, CA
Helen Kloepper	Louisberg Elementary and Middle School, Louisberg, KS
Bill Lescohier	Edna Maguire Elementary School, Mill Valley, CA
Kichung Lizee	Lac du Flambeau Public School, Lac du Flambeau, WI
Anne Marie Lord	Radnor Middle School, Wayne, PA
Carmon McFawn	Gillwinga Primary School, South Gafton, New South Wales, Australia
Trish Mihalek	St. Brigid School, San Francisco, CA
Colleen K. Morimoto	Waimanalo Elementary and Intermediate School, Waimanalo, HI
Robin CP Vernuccio	St. Ignatius Loyola School, New York
Suzanne Warmann	River Oaks Elementary School, Houston, TX
Robin Wexler	Roosevelt School, River Edge, NJ

Science Reviewers

Content Reviewed	Science Reviewer Affiliation
Mission 1: Your SETI Academy Medical File	Dr. Charles Wade—NASA-Ames, Moffett Field, CA
Mission 2: Using a Microscope	Dr. Friedmann Freund—SETI Institute, Mountain View, CA; NASA-Ames
Mission 3: Ancient Life-Forms	Dr. Lawrence Hochstein—NASA-Ames
Mission 4: Who Changed Earth's Atmosphere?	Dr. Linda Jahnke—NASA-Ames
Mission 5: Fossils!	Dr. David Milne—Evergreen State College, Olympia, WA
Mission 6: Natural Selection	Dr. David Milne—Evergreen State College
Mission 7: A Timeline for the Evolution of Life	Dr. David Des Marais—NASA-Ames
Mission 8: Tracing Family Trees	Dr. David Milne—Evergreen State College
Mission 9: What Organism Do You See?	Dr. Harold Klein—Santa Clara University, Santa Clara, CA; SETI Institute
Mission 10: Inventing Life-Forms	Dr. Dale Russell—Canadian Museum of Nature, Ottawa, Quebec
Mission 11: Creating Your Extraterrestrial's Family Tree	Dr. Dale Russell—Canadian Museum of Nature
Mission 12: Mission Complete!	None (no science content)

Special Acknowledgments

The SETI Institute Life in the Universe team thanks the following people for their help, inspiration, insights, support, and ideas contributed over the three-year period during which this project was developed.

Bob Arnold	Pam Bacon	Bernadine Barr	John Billingham
Linda Billings	Dave Brocker	Vera Buescher	Dawn Charles
David Chen	Tom Clausen	Gary Coulter	Kent Cullers
Seth DeLackner	Edna DeVore	Laurence Doyle	Frank Drake
Alice Foster	Friedmann Freund	Tom Gates	Janel Griewing
Sam Gulkis	Bud Hill	Wendy Horton	Garth Hull
Don Humphreys	Mike Klein	Carol Langbort	Steve Levin
Ivo Lindauer	Kathleen Marzano	Michelle Murray	Chris Neller
Barney Oliver	Ed Olsen	Frank Owens	Ray Reams
Don Reynolds	Hal Roey	Carol Stadum	

Introduction

Learning Objectives

Concepts

Through the activities in this book, students will learn about and be able to apply concepts in the following areas:

- Microscopes are an important tool for scientific investigations.

- Bacteria, algae, and protists are different forms of microscopic life that exist around us all the time.

- Throughout most of Earth's history, the only forms of life on Earth were microscopic.

- The oxygen in Earth's atmosphere that keeps us alive today was created by some microscopic organisms.

- Complex life-forms evolved from simpler forms. It is possible to trace these general developments in the fossil record.

- People evolved recently in comparison with other forms of life. The time periods over which the various life-forms evolved on Earth is immense in comparison with a single human lifetime.

- Factors that have shaped the evolution of life on Earth have included various environments and climates, catastrophic events that changed these environments, and chance.

- There are many similarities and differences among Earth plants and animals. These characteristics allow us to classify them and give clues as to how they evolved.

- Because environments on other planets are almost certain to be different from those on Earth, any life on other planets is almost certain to have evolved into different forms than it did on Earth.

Skills

The activities are also designed to help students develop the following abilities:

- Using a microscope.

- Arranging events in chronological order.

- Conducting, analyzing, and interpreting biochemistry experiments.

- Using the metric system.

- Visualizing vast expanses of time.

- Using a simple classification key.

- Synthesizing knowledge and creative imagination to visualize the evolution of life.

- Using simulations.

- Making scale models.

- Performing controlled experiments.

- Making laboratory measurements and analyzing them.

- Making graphs and interpolations from them.

- Creating maps.

Timeline and Planning Guide

Classroom trials suggest that this whole unit requires four weeks of activities, if science is taught every day. Time spent on the *SETI Academy Planet Project* activities can also be considered as time spent in mathematics and language arts, because these topics are emphasized right along with science, as part of an academically integrated program. Each mission subdivision is designed to take one class period. Some teachers may want to take two or even three class periods to teach some mission subdivisions.

Mission 1: Your SETI Academy Medical File

Students are introduced to the *SETI Academy Planet Project* as they meet the Academy's Chief Project Scientist and Executive Director, who ask students to consider that they may not be the only intelligent species in the universe. Students complete a medical profile for the SETI Academy.

Mission 2: Using a Microscope

Mission 2.1: Students become familiar with the parts of a microscope and watch a demonstration of how to operate a three-power microscope.

Mission 2.2: Students use a microscope to examine colored threads, a feather, newsprint, and rug fibers.

Mission 3: Ancient Life-Forms

Mission 3.1: Students use microscopes and hand lenses to observe some of the descendants of ancient organisms, including bacteria, cyanobacteria (blue-green bacteria), green algae, and planaria. Students imagine that they are investigating Earth 600 million years ago, drawing the life-forms they discover.

Mission 3.2: Students look over their drawings and discuss the questions, "What are the various ways that the life-forms you observed differ from each other?" and "Which life-form do you think evolved first, second, and so on?"

Mission 4: Who Changed Earth's Atmosphere?

Mission 4.1: Students perform an experiment to see which of the microscopic life-forms they studied in mission 3 give off oxygen.

Mission 4.2: Students comment on changes they see occurring in their experiment and begin to track down the "O_2 Culprit"; mission 4.2 should take place four to seven days after mission 4.1.

Mission 4.3: Students see a video image show that describes the formation of Earth and the early stages of the evolution of life on Earth.

Mission 5: Fossils!

Mission 5.1: Students create "fossils" from shells or plastic animals inside "sedimentary rock layers" made of dirt and sand as they make "Fossil Jars."

Mission 5.2: Students simulate the effects of mountain building, the creation of canyons, volcanism, and erosion upon sedimentary rock layers and the fossils within them.

Mission 6: Natural Selection

Students will become "hungry finches" as they play the "Hungry Finch Game" to demonstrate the process of natural selection.

Mission 7: A Timeline for the Evolution of Life

Mission 7.1: Students see a video image show that describes the later stages of the evolution of life on Earth.

Mission 7.2: In teams of two, students make a timeline that recaps all the major stages of the evolution of life on Earth using drawings from the video image show.

Mission 8: Tracing Family Trees

Mission 8.1: Students take part in a discussion about evolution and the work of paleontologists. They construct a recording sheet showing geologic periods.

Mission 8.2: Students arrange evidence from the fossil record to construct plausible family trees for trilobites and for the major vertebrate classes.

Mission 9: What Organism Do You See?

Mission 9.1: Students learn how a dichotomous classification key works as they divide 20 different objects into consecutively smaller groups.

Mission 9.2: Students play a game, What Organism Do You See?, that leads them through using the dichotomous classification key to identify different organisms.

Mission 10: Inventing Life-Forms

Students design an extraterrestrial life-form based upon characteristics that distinguish life-forms on Earth. Each student will use a combination of chance probabilities from dice rolling and creativity to imagine one such creature.

Mission 11: Creating Your Extraterrestrial's Family Tree

Students refer to the family trees they made in mission 8 to create family trees showing how their extraterrestrial creatures from mission 10 might have evolved from a worm-like organism.

Mission 12: Mission Completed!

Students write and draw what they have learned about the evolution of simple and complex life on Earth and the time involved in evolution.

Preparation

On Teaching About Evolution

These materials portray the evolutionary origin and development of life on Earth as understood by scientists, and direct students to apply those principles in their visualization of life on other planets. We recognize that the scientific evolutionary view conflicts with the religious views of some students. This conflict can interfere with learning, unless handled tactfully. If the issue comes up, it is helpful for all students to be assured that each person is entitled to their own individual belief systems, and that everyone is encouraged to talk about his or her opinions without fear of ridicule. It is also important to emphasize that your class is intended to help students *understand* scientific theories. For their personal comfort in class, students may honestly be assured that they need not subscribe to an evolutionary view to conduct these activities. They simply need to know what that view is.

Although biologists have seen enough evidence to convince them that life evolved, most students have not. As with every other scientific concept, it is important to encourage students to ask, "How do we know that it's true?" We can never "know" about events of the past, but we can gain increased confidence from observations that all converge toward the same interpretation. Observations from many scientific disciplines—geology, biochemistry, genetics, zoology, embryology, zoogeography—support the interpretation that life originated spontaneously and evolved the complexities that we see today. It is always conceivable that new evidence may be found that reverses or disproves any firmly established idea, including this one. In that sense, evolution is a "theory" about the past, individual aspects of which are potentially subject to disproof. It would be misleading to imply, however, that the "theory" of evolution is tentative and so unsupported by evidence that disproof by new evidence is likely. Intellectual honesty requires that students should be informed that evolution is so firmly substantiated by such a wealth of evidence that most biologists are completely confident that it explains the development of life on Earth.

SETI Institute and "SETI Academy"

The SETI Institute is a real scientific organization, but the SETI Academy is pure *fiction*. It is a device to increase student involvement in this material. The people who are listed in each mission as members of the "SETI Academy Team" are real scientists and science educators, but most of them have never met one another!

Assessment

The projects in this unit are designed to help teachers assess their students' understanding. The SETI Academy Cadet Logbooks will help teachers determine their students' grasp of the concepts and skills presented in the lessons, experiments, and projects. Use the student logbooks and the students' projects, along with student participation, to assign grades and to provide appropriate feedback. Read mission 12, "Mission Completed!," before copying any of its worksheets for the student logbooks. The questions on the "Mission Briefing" sheet can be used as a test, if desired, in which case students should not see them ahead of time.

Planning

Before teaching this book, teachers may wish to work through *The Evolution of a Planetary System*. When students have completed *How Might Life Evolve on Other Worlds?*, teachers may want to proceed directly into *The Rise of Intelligence and Culture* in which students endow a given life-form with intelligent capabilities and evolve cultures for that life-form. Each book is based on the detailed study of life as it has evolved on Earth and on the scientific observation of our neighborhood in space. Each book of the *SETI Academy Planet Project* contains an exciting, informative, and creative series of integrated science activities for upper elementary students.

Expanding or Compressing the Unit

There are a variety of activities that can add richness to this unit. Activities such as museum or zoo field trips and fossil hunts can add interest to the program. Concepts presented in each mission can be expanded upon by providing extra activities on the various topics presented in this unit. There are some ideas in the "Going Further" section in each mission. If short on time, consider cutting some of the missions or mission subdivisions less crucial for your class because of their background.

SETI Academy Cadet Logbooks

Masters for "mission briefings," student worksheets, and student handouts, all of which should be included in the student logbooks, are provided following the details of each mission. A master for the logbook cover is provided in mission 1, "Your SETI Academy Medical File." Also make a copy of the two-page glossary (found at the end of this book) for each student logbook. Paper is a limited resource both environmentally and at some school sites, so the following options for reproducing the masters are included.

Option One

Ideally, it is best to make copies of each master and assemble them into packets, one for each student. This option really captivates students by allowing each their own SETI Academy Cadet Logbook. This option involves reproducing about 50 pages. The reproduction can be done using a two-sided copier, but note that some of the student logbook pages are consumable, so be sure to copy those particular pages one-sided. When reproducing pages for the student logbooks, use three-hole-punched paper so students can keep their logbook papers in binders alongside other papers. Papers may be collated and handed out as complete logbooks or kept in folders to be handed out one or two sheets at a time as students are ready for new missions.

Option Two

Save on materials costs by producing one copy of the student logbook for every group of two or three students.

Option Three

For those schools that have a limited supply of paper, teachers might try making transparencies of the "mission briefing" masters and using them on an overhead projector. Save the transparencies and reuse them each year. Have students copy and answer the pre-activity "What Do You Think?" questions and the post-activity "What Do You Think, Now?" questions onto their own binder paper, which should be placed in their logbooks. Reproduce student worksheets and handouts from each mission and distribute them as needed.

Cooperative Learning

The *SETI Academy Planet Project* is well suited to the use of cooperative learning groups. Each group can have a materials monitor, a recorder, a speaker, and so forth. Successful cooperative learning groups should have a mix of learning styles and be balanced in sex and ethnicity. It is best if groups last at least several class periods so students have a chance to work together long enough to get comfortable. If two or all three of the *Project* books will be taught, new groups can be formed for each following unit so that students have opportunities to work with different peers.

Preparation of Special Materials

Video Images: History of Earth

A set of full-color images was produced by artist Jon Lomberg specifically for the Life in the Universe series. This set of images is used in several missions in *How Might Life Evolve on Other Worlds?*, as well as in missions in *The Evolution of a Planetary System*. For convenience, these images have been reproduced in a video format as the *History of Earth* for sale and as black-line masters reproduced in the back of this book. Each time the images are used, they are accompanied by different text—the "scripts" in the missions. Both the images and the scripts are printed in this teacher's guide.

Bulletin Board

Student groups will produce colorful, creative materials throughout this unit. Teachers will want to display their work somewhere in the classroom area. A lot of space will be required to display the students' work. If possible, set aside one entire wall of the classroom for this purpose.

Laboratory Materials

Various laboratory materials are needed throughout these activities. Refer to the "Materials" section at the beginning of each mission for a checklist. All supplies and laboratory materials for the entire unit are also on a composite "Required Materials List" in the appendixes of this book. Some items, such as the living cyanobacteria (blue-green bacteria) need to be ordered or cultured long before the unit begins.

Mission 1
Your SETI Academy Medical File

Overview

Jill Tarter and the other scientists students will be meeting throughout their missions at the SETI Academy are actual scientists affiliated with the SETI Project. The challenges they will pose to students illustrate the flavor, if not the complexity, of the kinds of things these scientists do.

The first mission is designed to be fun—it is intended to entice students into thinking that human beings may not be the only intelligent species in the universe. It is also meant to generate discussion about such intelligent species, on Earth and elsewhere.

Concepts

- Humans beings may not be the only intelligent species in the universe.

- Scientists are currently searching for extraterrestrial life by scanning the sky with radio telescopes.

Skills

- Analyzing similarities and differences.

- Critical thinking.

Mission 1

Materials

For Each Student

- SETI Academy Cadet Logbook

- Pencil

Getting Ready

One or More Days Before Class

1. Make copies of logbooks.

Just Before the Lesson

1. Assign pairs of students.

Classroom Action

1. **Introduction.** Tell students about the *SETI Academy Planet Project* and its emphasis on detecting extraterrestrial life. Point out that for most of the last 4 billion years, there was no intelligent life here on Earth capable of being detected; our species had yet to evolve. In the guide, we will investigate evolution on other worlds.

2. **Project Briefing**. Have the class refer to the "Project Briefing" from Chief Project Scientist Jill Tarter in their student logbooks while one student reads it aloud.

3. **Discussion**. Students may be interested to know that the SETI Institute is a real, non-profit corporation created to search for signs of intelligent life in the universe. The search activity was originally funded by the U.S. government; now searches by SETI Institute scientists are funded by private donations. Jill Tarter is a real scientist who has been involved in this project since 1975. Explain that the people in the photos in each mission are real scientists and science educators, though most of them have never met one another.

   The recent (1993) congressional decision to terminate funding for SETI searches, and the extreme disappointment expressed by the public, is an interesting example of how our representative form of government works. Some teachers have used this as an opportunity to, in cooperation with social science teachers, create an illustrative lesson that students can internalize. The SETI Institute has received numerous offers from students who want to hold bake sales or do other projects to "save SETI."

4. **Mission Briefing**. Have the class refer to the "Mission Briefing" for mission 1 in their student logbooks while one student reads it aloud.

5. **What Do You Think?** Have students answer the pre-activity questions on the "Mission Briefing." Invite them to share their answers in a class discussion.

6. **Activity**. Divide students into pairs and discuss the SETI Academy medical file "Questionnaire" logbook sheet. Each student should then fill out their own form. Tell students that completing the questionnaire is a way to start thinking about characteristics of species different than our own.

7. **Discussion**. Invite students to share their ideas and answers. Do not judge the students' ideas as right or wrong at this point. It is not important that they get the "right" answers for the questionnaire; rather, the idea is to get them to think about the similarities and differences among living creatures. Encourage discussion.

Ask the students to brainstorm about other intelligent species here on Earth. List these animals on the chalkboard, even if students disagree on any particular animal's "intelligence."

Invite students to describe the ways the named animals exhibit "intelligence." Create a chart; have students compare and contrast the animals' characteristics of intelligence. Encourage different definitions of the word *intelligence.*

Point out that the various animals named have very different body structures than humans. Ask students to describe some of these differences. *Some live in the water, others in trees, some are four-footed and have fur all over them.*

Stress that some scientists believe that, because our Sun is like many other stars in the Milky Way galaxy, it is possible that intelligent creatures may exist on planets around other stars. If so, such creatures would probably have very different body structures than humans. This is because the evolution of creatures depends on the environment and on chance variations in body types. It is almost impossible that there would exist another planet exactly like Earth, on which evolution took precisely the same course.

Discuss and define the term *evolution*. Tell students that they will find out in the next few missions how evolution proceeded on Earth.

Closure

1. **Discussion**. As an optional assignment class discussion, invite students to share their opinions about the "What do you think?" questions on the "Mission Briefing" logbook sheet

Going Further

Research: Smart Animals

Have students write reports about why certain species, such as pigs, dolphins, or whales, are considered more intelligent than other species.

Creative Writing: Once Upon a Time

Have students write creative stories about encounters with intelligent life-forms from another planet.

Activity: Truth and the Tabloids

Have students bring in articles from various tabloids that "prove" that there is life on other worlds. Have them look for articles on the "Great Stone Face" of Mars, cities on the Moon, and so forth. Ask students how to determine whether or not such stories are truthful. Some students may believe anything that they read, while others will automatically disbelieve anything that appears in a tabloid, whether it is true or not. Try to develop logical strategies for analyzing stories from tabloids or from the daily newspaper. Encourage students to be skeptics!

Arrange a lesson on how easy is it to "fake" photos, especially given the computer graphics technology available today. Many tabloid claims are simply hoaxes. Challenge students to make their own convincing "hoax" photos or videos of UFOs and ETs!

Ask students what sort of proof would be enough for them to believe an extravagant claim, such as a flying saucer landing on Earth. Such a "grand" claim would require "grand" evidence. A physical artifact (not a photo of one) that could be tested to prove that it could not possibly have been made on Earth would be such evidence. Or, a new fact, one not previously known to anyone on Earth, that could be substantiated would be "grand" evidence.

Discussion: Funding SETI

For many years, the U.S. government funded much of the Search for Extraterrestrial Intelligence. But late in 1993, Congress cut all funding. Currently, SETI is being supported entirely by private donations. Discuss whether the government should fund SETI.

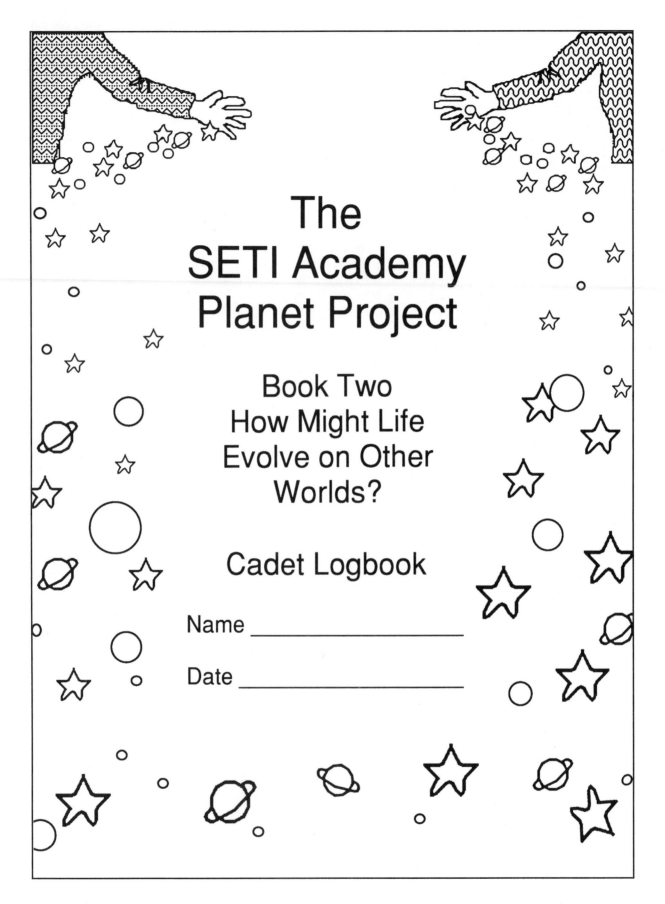

The SETI Academy Planet Project

Book Two
How Might Life Evolve on Other Worlds?

Cadet Logbook

Name _____

Date _____

Mission 1

Planet Project Briefing

Name:

Date:

Dr. Jill Tarter, Chief Project Scientist of the SETI Academy.

Welcome to SETI Academy, a training program for students interested in space sciences. As you may know, I work with other scientists on the Search for Extraterrestrial Intelligence (SETI). Our job is to search for signs of intelligent life in our Milky Way Galaxy. Our home base is in Mountain View, California, where we work closely with scientists from the National Aeronautics and Space Administration (NASA) at Ames Research Center. Of course, if there are other intelligent beings in our galaxy, we cannot know for sure what they or the other inhabitants of their planets are like until we discover them. But, as scientists, we can still make some intelligent guesses.

Your task is to simulate how life may evolve on other planets. Start by learning as much as possible about early microscopic life on Earth and how it evolved into humans, whales, dolphins, dogs, cats, and other intelligent species. Then, apply what you learn to inventing a reasonable simulation of how complex creatures might have evolved on some other planet orbiting a distant star. Good luck!

Mission 1

Your SETI Academy Medical File
Mission Briefing

Name:

Date:

Tom Pierson, Executive Director of the SETI Academy.

If we should someday discover other intelligent beings who live on other planets in the Milky Way Galaxy, we will certainly want to invite them to join SETI Academy. In the meantime, we are creating a questionnaire to help us construct a medical profile of each SETI Cadet. The questionnaire will tell us all about his, her, or its species. We hope that you and your classmates will join the SETI Academy to become the next class of SETI Cadets. Please help us by filling out this SETI medical questionnaire. You can be an Earth species (a human being or an animal) or an imagined extraterrestrial species. When you finish the questionnaire, answer the following questions.

What Do You Think?

1. Which intelligent species besides humans share our planet? Choose one such animal and imagine it can read and write (if you answered the questionnaire as if you were an animal, choose a *different* animal). If this animal completed the questionnaire, which answers would be different from yours?

2. Do you think any other Earth species should be invited to join SETI Academy now? Why or why not?

3. If we discover intelligent beings from other planets, how will we *know* that they are intelligent?

From *How Might Life Evolve on Other Worlds?* © 1995. Teacher Ideas Press. (800) 237-6124.

Mission 1

Your SETI Academy Medical File Questionnaire

Name:

Date:

Your Age:

Please answer by checking the boxes. Some questions can have more than one answer.

1. What are you?

☐ a plant
☐ an animal
☐ neither
☐ other:_____
☐ don't know

2. What parts of your body are hard?

☐ an inside skeleton
☐ an outside shell
☐ no hard parts
☐ other:_____
☐ don't know

3. What covers your body?

☐ leathery hide
☐ smooth skin with fine hairs
☐ scales
☐ other:_____
☐ don't know

4. What is your body shape?

☐ longer than it is wide
☐ round
☐ other:_____
☐ don't know

5. How much do you weigh?

- ☐ 0-50 lbs.
- ☐ 51-200 lbs.
- ☐ 201-500 lbs.
- ☐ other: _____
- ☐ don't know

6. Does your body have sections?

- ☐ simple, attached segments that are all identical (like a centipede)
- ☐ a head, a thorax, and an abdomen (like an insect)
- ☐ a head and a body
- ☐ other: _____
- ☐ don't know

7. How many appendages (arms, legs, or tentacles) do you have?

- ☐ 2
- ☐ 4
- ☐ 5
- ☐ other: _____
- ☐ don't know

8. How many body openings do you have?

- ☐ 0
- ☐ 1-5
- ☐ 6-10
- ☐ other: _____
- ☐ don't know

9. What do you eat?

- ☐ meat
- ☐ junk food
- ☐ plants
- ☐ other: _____
- ☐ don't know

10. What sense organs do you have?

- ☐ ears
- ☐ nose
- ☐ eyes
- ☐ skin
- ☐ tongue
- ☐ antennae
- ☐ hands
- ☐ other: _____
- ☐ don't know

11. What do your sense organs detect?

☐ vibrations
☐ visible light
☐ chemicals
☐ heat
☐ other:_____
☐ don't know

12. How does your body get food (nutrients) and oxygen to its cells?

☐ through its blood
☐ directly from its environment
☐ other:_____
☐ don't know

13. If your body has blood, how is it circulated?

☐ with one heart or blood pump
☐ with two or more hearts or blood pumps
☐ other:_____
☐ don't know

14. How do you protect yourself from enemies?

☐ hiding
☐ working with others
☐ running away
☐ biting
☐ freezing in your tracks
☐ other:_____
☐ don't know

15. How does your body stay warm?

☐ warm-blooded, creates its own heat
☐ cold-blooded, gets heat from environment
☐ other:_____
☐ don't know

16. How does your species reproduce?

☐ sexually
☐ asexually
☐ other:_____
☐ don't know

Mission 2

Using a Microscope
The World of the Very, Very Small

Overview

In mission 2.1, students become familiar with the parts of a microscope and watch a demonstration of how to operate a three-objective microscope. Effective use of microscopes is an important skill for scientific inquiry in many disciplines of science. This mission is provided for students who have limited or no experience using microscopes. In mission 2.2, students use a microscope to examine colored threads, a feather, newsprint, and rug fibers.

Concepts

- Microscopes are an important tool for scientific investigations.

Skills

- Using a microscope.

- Analyzing specimens.

- Using laboratory techniques.

Mission 2.1

Materials

For the Class

- 3 Transparency Mask masters (pages 28-30)

- Overhead projector

- Transparency of the "Microscope Transparency" sheet (see page 27)

Notes

In mission 1, students were briefed about the SETI Academy Planet Project and introduced to the Chief Project Scientist and Executive Director of the real SETI Institute. They completed a SETI Academy medical questionnaire.

For Each Student

- SETI Academy Cadet Logbook

- Pencil

Getting Ready

One or More Days Before Class

1. Using the three "Transparency Mask" masters, make masks for the overhead projector by cutting 6-inch, 4-inch, and 2-inch diameter circles from the masters supplied.

2. Make an overhead transparency of the "Microscope Transparency" sheet (page 27).

Just Before the Lesson

1. Set up the overhead projector. Test the masks and the transparency. Make sure there is a clear path for moving the overhead projector backward to positions that keep the diameter of the circle approximately the same for each mask, as shown in figure 2.1.

Figure 2.1—Overheads with Masks.

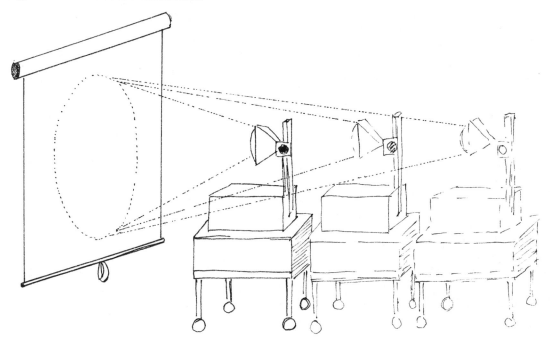

Classroom Action

1. **Mission Briefing**. Have the class refer to the "Mission Briefing" for mission 2 in their student logbooks while one student reads it aloud.

2. **What Do You Think?** Have students answer the pre-activity questions on the "Mission Briefing." Invite them to share their answers in a class discussion.

3. **Lecture**. Tell students that they will be learning to use a microscope to help them in their search about how life might have evolved on Earth, and that they will be applying that knowledge to the possible evolution of life on other worlds.

4. **Demonstration**. Familiarize students with the parts of a microscope. Remember to stress that one should never touch the surface of the lenses, the cover slips, or the slides; even clean fingers leave greasy residues. Stress that one should never twist the focusing knob toward the slide and cause the lens to touch the slide; high-power objectives can smash through slides, causing expensive damage. Stress that one should always pick up and carry the microscope by the arm.

   Give students a demonstration of how to operate and focus a three-power microscope. Place the transparency of the microscope on your overhead projector. Because microscopes invert the image, put the transparency on the projector upside down. Point this out to students. Tell students that each circle represents the view through a different-magnification eyepiece of the microscope.
   Place the 6-inch mask on the transparency. Focus the image on the screen. Tell students that the knob being used to focus the transparency image represents the knob used to focus a microscope. Discuss the projected transparency image. Tell students that what they are seeing is similar to what they would see when looking at an organism through the lowest-magnification lens on the microscope. Discuss the features that can be seen at this magnification.

Next, put the 4-inch mask on the transparency and push the overhead backward a few feet and refocus the image of the microscope on the screen. Explain to students that changing the microscope to a higher power of magnification gives a closer view of a smaller area.

Repeat this with the 2-inch mask, which represents the highest power of magnification. With each mask, discuss how the view includes more detail but less of the whole specimen. At the final 2-inch "magnification," point out that it is now possible to read the word "High" on the knob on the stage of the large microscope, as shown in figure 2.2.

Figure 2.2—Parts of a Microscope.

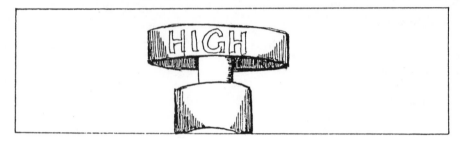

Mission 2.2

Materials

For the Class

- 4 to 16 three-objective microscopes, student quality

- 4 to 16 plastic or glass slides

- 4 to 16 plastic cover slips

- 12 1-cm squares of newspaper articles

- 3 spools of thread, contrasting colors

- Feather

- Samples of fiber from shag-type carpets

- 1 to 4 trays for center supplies

- Copies of the "Microscope Center Directions" logbook sheets

- Copies of the "Parts of a Microscope" logbook sheet.

For Each Student

• SETI Academy Cadet Logbook

• Pencil

• Colored pencils or crayons

Getting Ready

One or More Days Before Class

1. Locate microscopes and related supplies. The quality and power of these microscopes will vary. Decide how many Microscope Centers can be set up. Cut up a feather into as many pieces as there will be Microscope Centers.

2. Locate an assistant or volunteer, if possible, who can circulate among the students to help them focus the microscopes.

3. Make one copy of the "Microscope Center Directions" logbook sheet and one copy of the appropriate "Parts of a Microscope" logbook sheet for each Microscope Center.

Just Before the Lesson

1. Set up Microscope Centers with the microscopes. Make sure each center has a microscope and a tray with the necessary supplies. Don't forget to have towels, buckets of water, and trash cans for cleanup at each Microscope Center. The following are suggestions for organizing the classroom:

   16 microscopes. If you have 16 microscopes, set up the classroom in the ideal situation shown in figure 2.3. Place supply tray(s) near the appropriate microscopes. Teams of four to eight students each begin at a table, set up a slide, and record what they see. Then the teams travel from table to table, viewing and recording what they see on their "Microscope Center Recording Sheets." It does not matter at which Microscope Center a student starts.

   4 microscopes. If you have only four microscopes, set up just one Microscope Center, with all supplies on one tray, and rotate the students through. (Refer to mission 3, Fig. 3.2, page 41 for an example.) Or, set up four stations with one microscope at each, then have half the class rotate through.

Single-objective microscopes. If you only have access to single-powered microscopes, use those in addition to the naked eye and a magnifying lens to achieve the three levels of magnification.

Figure 2.3—Organizing the Classroom: Option One 16 Microscopes.

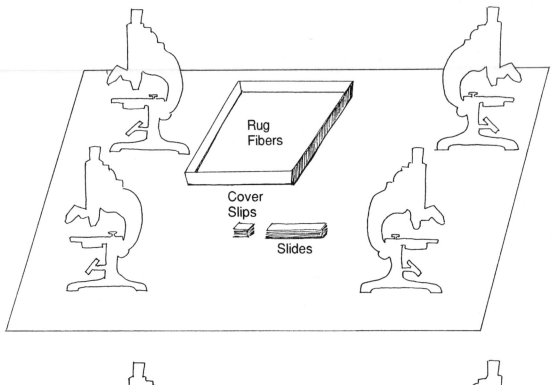

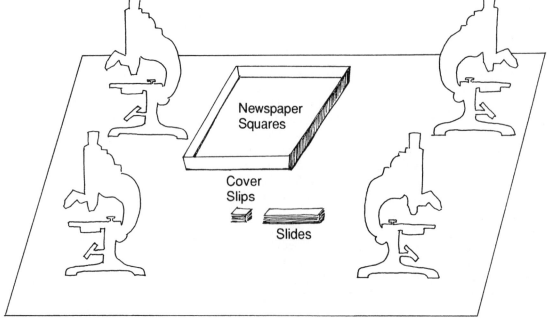

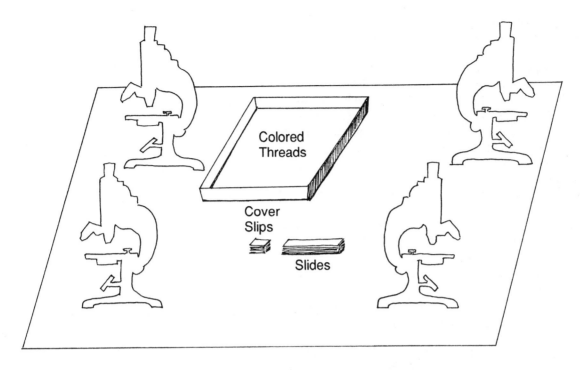

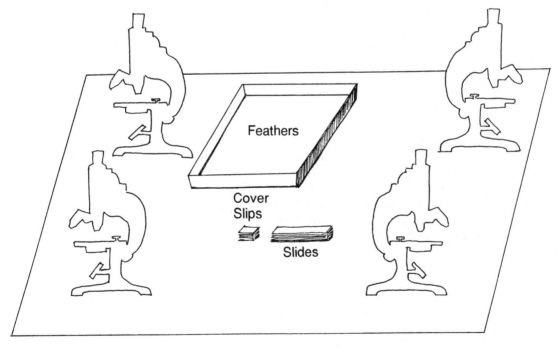

Classroom Action

1. **Activity.** Review the rules for proper use of the microscope. Also review the "Student Center Recording Sheets" for the Microscope Centers and present rules for behavior during this microscope lab. Allow time for students to complete each Microscope Center.

Closure

1. **Discussion**. After students complete the microscope lab, discuss the following questions.

   What was the most interesting part of the lab?
   Which specimen took the most skill to see when operating the microscope? Why?

2. **What Do You Think, Now?** Have students answer the post-activity questions on the logbook sheet "What Do You Think, Now?" Invite students to share their responses. Ask students how their opinions have been changed by this mission.

Going Further

Activity: What's on My Finger?

Have students put a piece of transparent tape on the palm of their hand or on the ventral side of their index finger after several hours of not washing. Make a slide by placing the tape over a 1-cm hole cut from the center of a slide-sized piece of index card.

Activity: What's in My Mouth?

The lab samples used in this mission are all viewed dry. This activity gives students practice with wet slides and gets students to realize how small cells are.

Have students use toothpicks to gently scrape the inside of their cheek. They will see some of their own cells! A drop of dye, like ink or food color, may help make this clearer. Have students scrape in between their teeth. A fairly common resident of the human mouth is a *Entamoeba gingivalis* amoeba; they may find one! Bacteria will be present, but most are too small to see well.

Activity: More Microscopic Stuff!

Try commercially prepared microscope slides of lizard skin, cat hair, and many other things. Or, have students make slides. Have them bring in hair samples from pets, or a dead butterfly: The scales on a butterfly's wing are beautiful. Devote one extra day to discovering the "Microworld."

Mission 2

Using a Microscope

Microscope Center #1 Directions

See what happens to the letter *e* when you put it under the microscope.

1. Select or cut out a 1-cm square of newspaper. Make sure the letter *e* is visible in the text. Place it on a slide as shown below, and carefully set a slip cover on top of the paper square.

Figure 2.4—Slide with Newspaper.

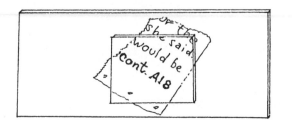

2. Gently put the slide on the stage of the microscope and ease a stage clip over each end of the slide to hold it in place. Turn the focusing knob to lower the tube of the microscope down toward the slide. (Make sure it doesn't touch the slide!)

3. Turn on the light under the stage of the microscope or position the mirror so it reflects light up through the slide. Rotate the lens piece until the lowest-powered lens clicks into place over the slide. Look through the eyepiece with one eye and close your other eye. You should see a blurry image of part of the newspaper square. Turn the focusing knob until the image becomes clear. Draw what you see in the appropriate circle on your recording sheet.

4. Rotate the lens piece until the middle-powered lens is over the slide and refocus the image as you look through the eyepiece. Draw what you see in the appropriate circle on your recording sheet.

5. Rotate the lens piece until the highest-powered lens is over the slide and refocus the image as you look through the eyepiece. Draw what you see in the appropriate circle on your recording sheet.

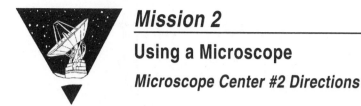

Mission 2

Using a Microscope

Microscope Center #2 Directions

Examine threads that are crisscrossed to learn focusing techniques at different depths.

1. Select three pieces of thread, different colors. Place them on a slide as shown below, and carefully set a slip cover on top of the crisscrossed threads.

Figure 2.5—Slide with Colored Threads.

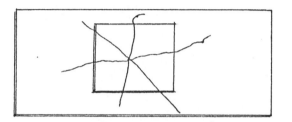

2. Gently put the slide on the stage of the microscope and ease a stage clip over each end of the slide to hold it in place. Turn the focusing knob to lower the tube of the microscope down toward the slide. (Make sure it doesn't touch the slide!)

3. Turn on the light under the stage of the microscope or position the mirror so it reflects light up through the slide. Rotate the lens piece until the lowest-powered lens clicks into place over the slide. Look through the eyepiece with one eye and close your other eye. You should see a blurry image of the colored threads. Turn the focusing knob until the image becomes clear. Draw what you see in the appropriate circle on your recording sheet.

4. Rotate the lens piece until the middle-powered lens is over the slide and refocus the image as you look through the eyepiece. Draw what you see in the appropriate circle on your recording sheet.

5. Rotate the lens piece until the highest-powered lens is over the slide and refocus the image as you look through the eyepiece. Draw what you see in the appropriate circle on your recording sheet.

Mission 2

Using a Microscope

Microscope Center #3 Directions

Examine twisted rug fibers to learn focusing techniques at different depths.

1. Choose a strand of shag-type carpet fiber from the tray and shred it apart to flatten it slightly. Place it on a slide as shown below, and carefully set a slip cover on top of the flattened fibers.

Figure 2.6—Slide with Rug Fibers.

2. Gently put the slide on the stage of the microscope and ease a stage clip over each end of the slide to hold it in place. Turn the focusing knob to lower the tube of the microscope down toward the slide. (Make sure it doesn't touch the slide!)

3. Turn on the light under the stage of the microscope or position the mirror so it reflects light up through the slide. Rotate the lens piece until the lowest-powered lens clicks into place over the slide. Look through the eyepiece with one eye and close your other eye. You should see a blurry image of the carpet fibers. Turn the focusing knob until the image becomes clear. Draw what you see in the appropriate circle on your recording sheet.

4. Rotate the lens piece until the middle-powered lens is over the slide and refocus the image as you look through the eyepiece. Draw what you see in the appropriate circle on your recording sheet.

5. Rotate the lens piece until the highest-powered lens is over the slide and refocus the image as you look through the eyepiece. Draw what you see in the appropriate circle on your recording sheet.

Mission 2

Using a Microscope

Microscope Center #4 Directions

See the details that are visible in a feather when you put it under the microscope.

1. Choose a piece of feather from the tray and pull it apart to slightly separate the barbs. Place it on a slide as shown below, and carefully set a slip cover on top of the feather specimen.

Figure 2.7—Slide with Feather.

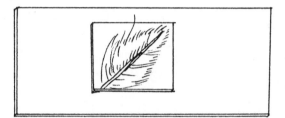

2. Gently put the slide on the stage of the microscope and ease a stage clip over each end of the slide to hold it in place. Turn the focusing knob to lower the tube of the microscope down toward the slide. (Make sure it doesn't touch the slide!)

3. Turn on the light under the stage of the microscope or position the mirror so it reflects light up through the slide. Rotate the lens piece until the lowest-powered lens clicks into place over the slide. Look through the eyepiece with one eye and close your other eye. You should see a blurry image of the feather. Turn the focusing knob until the image becomes clear. Draw what you see in the appropriate circle on your recording sheet.

4. Rotate the lens piece until the middle-powered lens is over the slide and refocus the image as you look through the eyepiece. Draw what you see in the appropriate circle on your recording sheet.

5. Rotate the lens piece until the highest-powered lens is over the slide and refocus the image as you look through the eyepiece. Draw what you see in the appropriate circle on your recording sheet.

Mission 2

Using a Microscope

Parts of a Microscope

Figure 2.8—Parts of a Microscope.

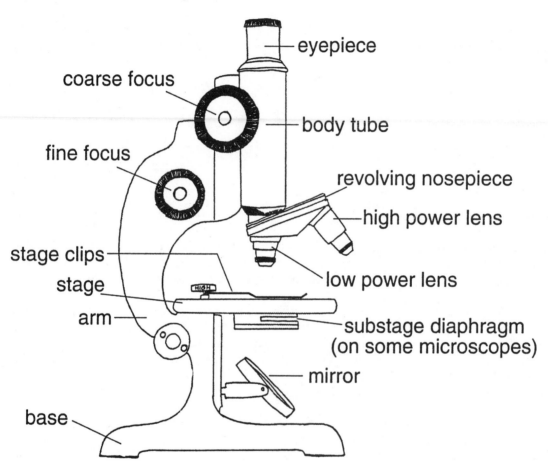

Mission 2

Using a Microscope

Microscope Transparency

(***Teacher's Note:*** *not a logbook sheet*)

Figure 2.9—Using a Microscope.

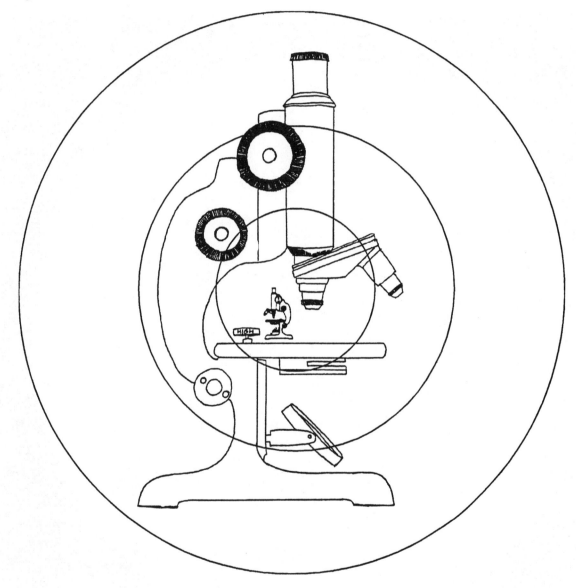

Mission 2

Using a Microscope

6-inch Transparency Mask

(***Teacher's Note:*** *not a logbook sheet*)

Figure 2.10—Six-Inch Transparency Mask.

Mission 2
Using a Microscope
4-inch Transparency Mask

(***Teacher's Note:*** *not a logbook sheet*)

Figure 2.11—Four-Inch Transparency Mask.

From *How Might Life Evolve on Other Worlds?* © 1995. Teacher Ideas Press. (800) 237-6124.

Mission 2

Using a Microscope

2-inch Transparency Mask

(***Teacher's Note:*** *not a logbook sheet*)

Figure 2.12—Two-Inch Transparency Mask.

Mission 2

Using a Microscope
Mission Briefing

Name:

Date:

**Lisa Chen, Research Assistant
on the SETI Academy Team.**

I have been informed by other scientists here at the SETI Academy that some of you have not had experience using microscopes. If this is the case with your class, please take advantage of this mission to become familiar with using a microscope to observe details of everyday items or organisms that are not visible to the naked eye.

What Do You Think?

1. What would a bird's feather look like under a microscope? Draw your idea.

2. What would the letter *e* in a newspaper look like under a microscope? Draw your idea.

3. Do you think that you could tell if something was on top of something else on a microscope slide? How?

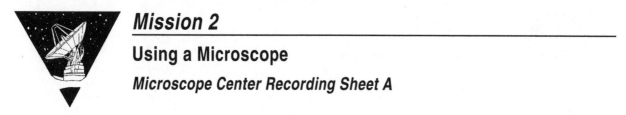

Mission 2

Using a Microscope

Microscope Center Recording Sheet A

Name:

Date:

Carefully draw what you see of the specimen at different powers in the circles below.

Figure 2.14—Newspaper Specimen.

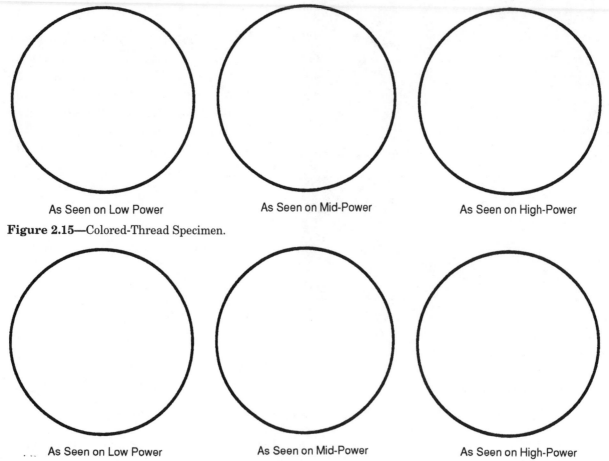

As Seen on Low Power   As Seen on Mid-Power   As Seen on High-Power

Figure 2.15—Colored-Thread Specimen.

As Seen on Low Power   As Seen on Mid-Power   As Seen on High-Power

Mission 2

Using a Microscope

Microscope Center Recording Sheet B

Name:

Date:

Carefully draw what you see of the specimen at different powers in the circles below.

Figure 2.16—Carpet-Fiber Specimen.

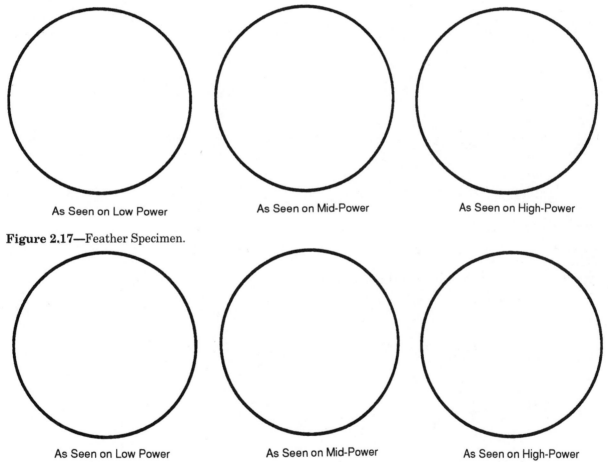

As Seen on Low Power	As Seen on Mid-Power	As Seen on High-Power

Figure 2.17—Feather Specimen.

As Seen on Low Power	As Seen on Mid-Power	As Seen on High-Power

Mission 2

Using a Microscope

What Do You Think, Now?

Name:

Date:

After you have completed this mission, please answer the following questions:

1. How can the microscope be used as a scientific tool?

2. What can you see with the microscope that you can't see with your unaided eyes?

Mission 3

Ancient Life-Forms
Earth, 600 Million Years Ago

Overview

In mission 3.1, students use microscopes and hand lenses to observe living microscopic organisms that are *like* the ancient life-forms that first appeared on Earth. Until about 730 million years ago, such creatures were the most common life on planet Earth. In mission 3.2, students hypothesize about which microscopic organisms evolved first.

For students to understand how life might evolve on other planets, they need to learn how life evolved on Earth. Although no one knows for certain how life on Earth began, there is ample fossil evidence that life began almost as soon as Earth formed a cool crust and a shallow ocean, some 3.8 billion years ago. The first forms of life were microscopic in size. The first multicellular organisms were also very small. Descendants of these organisms, organisms that are similar (but not identical) to these kinds of early life-forms, still survive today.

Concepts

- Bacteria, algae, and protists are different forms of microscopic life that exist all around us today.

- Throughout most of Earth's history, the only forms of life on Earth were microscopic in size.

- There is ample fossil evidence that life began to evolve almost as soon as Earth formed a cool crust and a shallow ocean, some 3.8 billion years ago.

- Some of today's living organisms resemble some fossil organisms; they are the same *kind* of organism, but are different species.

Notes

In mission 2, students became familiar with using a microscope.

Skills

- Using a microscope.

- Using laboratory techniques.

Mission 3.1

Materials

For the Class

- 4 to 16 microscopes

- 4 to 16 hand lenses

- Bacteria culture (grown on a potato)

- Culture of cyanobacteria (blue-green bacteria), 1/2 cup

- Culture of green algae, 1/2 cup

- Culture of protists

- Black or brown planaria

- 6 medicine droppers

- 16 to 24 plastic or glass slides

- 16 to 24 plastic cover slips

- 1 to 4 culture dishes (or shallow containers)

- Planaria food (fresh beef liver or hard-boiled egg yolk)

- Bean seeds or uncooked rice (small amount)

- 4 to 8 buckets

- Dish soap

- 1 to 4 towels

- 1 to 4 sponges

- Copies of "Microscope Center Directions" log-book sheets

- (optional) Pond water for microorganisms

- (optional) Timothy hay (dry grass) for microorganisms

- (optional) Specimen images (if you have no microscopes)

- (optional) Videocassette, VCR, and monitor

For Each Student

- SETI Academy Cadet Logbook

- Pencil

- Colored pencils or crayons

Getting Ready

Long Before Class

1. Plan out how to obtain the required organisms. See the appendixes for directions on how to grow and care for specimens as well as for ordering information.

One or More Days Before Class

1. This mission requires quite a bit of time for setup. An assistant, volunteer parent, or high school student would help reduce the time required.

2. The length of time students need for success at each station varies. An assistant, volunteer parent, or high school student available at each station during the class would help all students to see everything.

3. Choose one option for organizing the classroom.

   Demonstration microscope(s). For classrooms without enough equipment, set up one central materials location and place the organisms on slides as students request. Make sure they look at all of the specimens, in any order.

   Videotape. If you have no access to microscopes, this activity can be done using a videocassette segment showing four images of modern versions of ancient organisms or using an overhead projector and transparencies of the organisms.

16 microscopes. If you have 16 microscopes, set up the classroom as shown in figure 3.1. Place magnifiers, microscopes, eye droppers, slides, cover slips, culture dishes, and containers with pond water near the appropriate microscopes. Teams of four to eight students each begin at a table and set up a slide or feed planaria, and then record what they see. Then the teams travel from table to table viewing the organisms and recording what they see into their logbooks. For this option, 16 slides, 16 cover slips, 16 microscopes, and 4 hand lenses are needed. Set up a cleanup center for all students to use, consisting of a bucket of soapy water, a bucket of clear water, a towel, and a trash can.

Figure 3.1—Organizing the Classroom: 16 Microscopes.

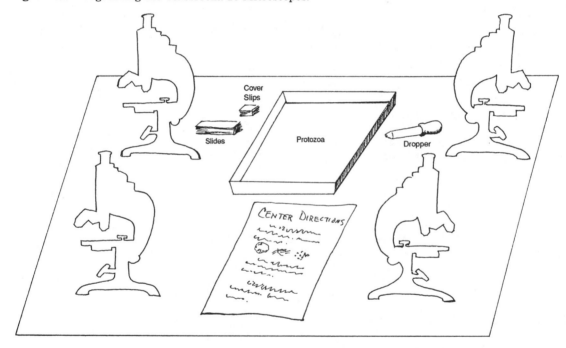

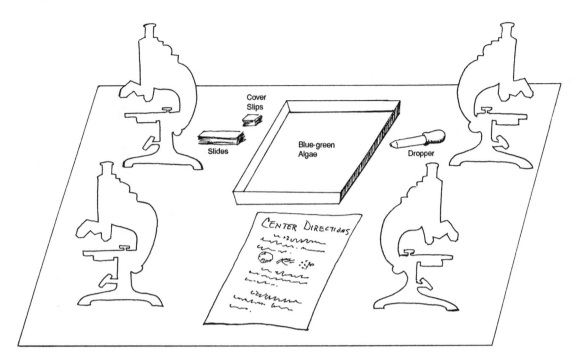

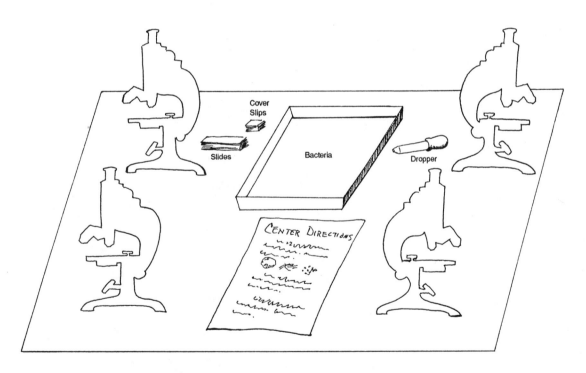

4 microscopes. If you do not have 16 microscopes, this session can be organized as a center for 10 students at a time, working in pairs. This option works well. Each student sets up one slide or feeds a planaria. Then the students rotate through the center, taking turns viewing the organisms and recording what they see into their logbooks. For this option, 24 slides, 24 cover slips, 4 microscopes, and 2 hand lenses are needed. Set up a cleanup center for all students to use, consisting of a bucket of soapy water, a bucket of clear water, a towel, and a trash can.

Figure 3.2—Organizing the Classroom: 4 Microscopes.

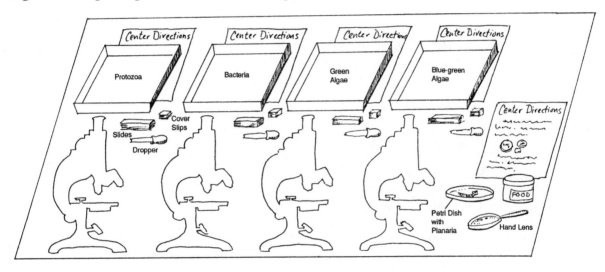

Just Before the Lesson

1. When preparing the microscope centers, put out the minimum number of organisms. Set aside the remainder for the next activity.

2. Make test slides and observe results under the microscope to be certain that the cultures are healthy.

Classroom Action

1. **Mission Briefing**. Have the class refer to the "Mission Briefing" for mission 3 in their student logbooks while one student reads it aloud.

2. **What Do You Think?** Have students answer the pre-activity questions on the "Mission Briefing." Invite them to share their answers in a class discussion.

3. **Discussion**. Microscopes and hand lenses will be used to observe the descendants of ancient organisms. Invite questions. Ask students to look at the "Microscope Center Recording Sheets" in their logbooks and notice the areas where they will record their observations.

    (optional) Show images or transparencies if microscopes are not available. View images or transparencies of protists, bacteria, cyanobacteria (blue-green bacteria), and green algae. Leave each image on the screen long enough for students to record the image as seen into their logbooks. Tell students to draw their pictures in the circles next to the words "view through microscope."

4. **Demonstration**. Ask students to gather around a microscope as you demonstrate how to make a wet-mounted slide, as shown in figure 3.3. Show how to select a *small* portion of an organism to be viewed.

    Bacteria and protists. Use an eyedropper to suck up a little of the water. Plenty of organisms should be in each drop. Put a tiny drop on the center of the slide. If, in your personal test of the microscope, you found the bacteria and protists to be very active, provide small pieces of a loose-weave fabric, such as cheesecloth, to put on the slide. This will help to trap the organisms and keep them from moving out of view. Set a cover slip on one edge of the drop of water and slowly lower it to the slide. Avoid trapping air bubbles.

Green algae and cyanobacteria (blue-green bacteria). Gently break off a tiny part of the cell mass. Put it in on the slide. Use the medicine dropper to move a drop of water from the container the organisms came from to the slide. Set a cover slip on one edge of the drop of water and slowly lower it to the slide. Avoid trapping air bubbles.

Planaria. Do *not* use the medicine dropper. The planaria can be viewed with a hand lens in the Petri dishes. One student may put food (a pinhead-sized piece of fresh beef liver or hard-boiled egg yolk) into the Petri dish. Other students should not add food until each "meal" has been eaten. Planarians are fragile. Instruct students not to touch them.

Figure 3.3—How to Prepare a Slide.

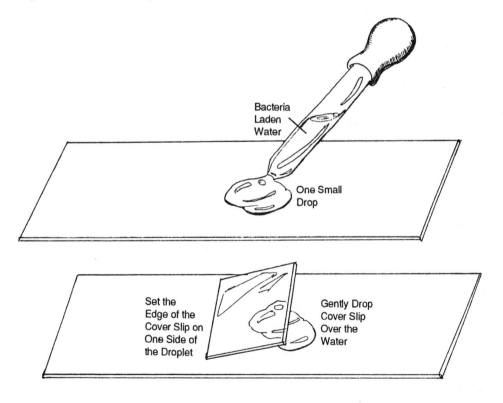

Show students how to place the slide on the microscope stage, how to arrange the light, and how to focus. Show students how to move the slide around until they have the sharpest, clearest view of the organism before drawing it.

Demonstrate how to wash and dry the slides and cover slips. Before washing slides, use the medicine dropper to put the water on the slide back into the water it came from. *Do not mix the organisms.* Continue handling the slides and cover slips by the edges and gently put them into the soapy water. Clean them off with a paper towel, slosh them through the fresh water, and lay them out to dry on the towel.

Go over some safety rules: No running. Be careful around glass. Do not touch the glass part of the eyepiece or the lens on the microscope. Use only the knobs to change focus or power. Do not touch the organisms.

Remind students that the point of this lesson is scientific exploration of life-forms that are in many ways like those that existed on Earth 3.8 billion years ago to the present. Until about 730 million years ago, such microscopic creatures were the *only* life on Planet Earth. They should find as many organisms as possible in the pond water and draw a clear, detailed view of each organism.

5. **Activity**. Have students use the microscopes. Allow extra time at stations for students to observe the findings of others, especially when there are a variety of protists.

Teacher's Note: *Most bacteria are much smaller than protists, green algae, and blue-green bacteria. Students will see only little dots unless they have at least 1,000X magnification on their microscopes.*

Mission 3.2

Materials

For Each Student

- SETI Academy Cadet Logbook

- Pencil

Getting Ready

No advance preparations are needed.

Classroom Action

1. **Activity**. Ask students to meet with their lab partners to share their drawings and to consider the questions asked in mission 3 What do you think, now? (see page 53).

2. **Discussion**. Hold a class discussion about which organism evolved first, which evolved second, third, fourth, and fifth during the 3.8 bya (billion years ago) to 730 mya (million years ago) period. If students are unfamiliar with the terms *bya* and *mya*, spend time on them now.

Encourage alternative ideas and challenge the students to come up with a sensible hypothesis (educated guess) to explain why they put the organisms in the order they did. For example, some students might say that plants evolved before animals, or that small life-forms evolved before bigger ones. Encourage discussion. Tell students that later they will find out what fossil evidence has to say about which of the life-forms is oldest. At this point, they should see that the planaria is clearly the most complex form; so it is reasonable to deduce that it is the organism that appeared most recently.

Teacher's Note: *The correct order is: bacteria (single-celled, without nuclei); cyanobacteria (single-celled, without nuclei); green algae and protists (single-celled, with visible nuclei) evolved at the same time; and planaria (a true multicellular animal; an invertebrate life-form).*

Closure

1. **What Do You Think?** Have students answer the pre-activity questions on the "Mission Briefing." Invite them to share their answers in a class discussion.

2. **Preview**. Tell students that, in the next mission, they will conduct laboratory experiments to help determine which of these organisms unintentionally caused a catastrophe that forced many other life-forms into extinction over two billion years ago!

Going Further

Creative Writing: Micro-Stories

Have students write adventure stories from the point of view of a microorganism.

Activity: A Micro-Safari

Have students go on a microbe safari, hunting for "trophy" microorganisms. They may bring in pond water or water from an aquarium. Suggest local places where they can hunt.

Activity: What *Is* That Thing?

Try to identify mixed protists or algae by using a field guide or an identification key. Several are on the market. A local high school or science center may have copies to loan.

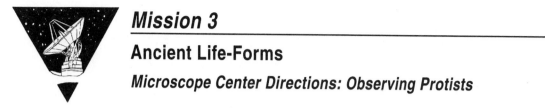

Mission 3

Ancient Life-Forms

Microscope Center Directions: Observing Protists

Figure 3.4—What to Look For.

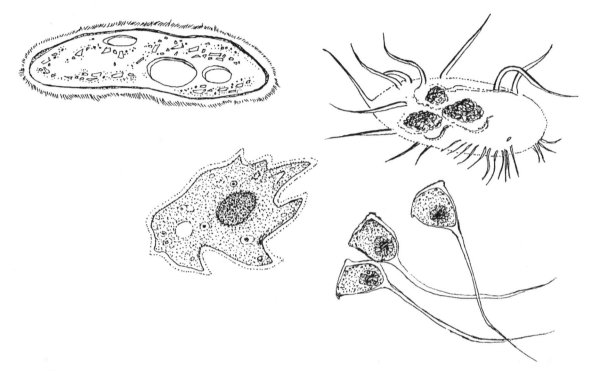

Protists are living organisms! Handle with care and *keep them wet*. If there is a slide already prepared, make sure the slide is still moist. If not, add one drop of water to the edge of the cover slip, as shown below.

Figure 3.5—Adding Moisture to a Slide.

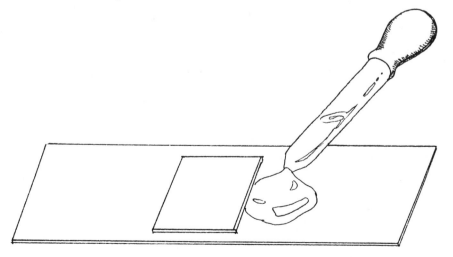

From *How Might Life Evolve on Other Worlds?* © 1995. Teacher Ideas Press. (800) 237-6124.

If you need to prepare a slide, cut a small piece (smaller than a cover slip) of loose-weave fabric and place it on the center of the slide. Place a small drop of water with protists on the fabric, as shown below, and cover with a slip.

Figure 3.6—Preparing a Wet Mount with Loose-Weave Fabric.

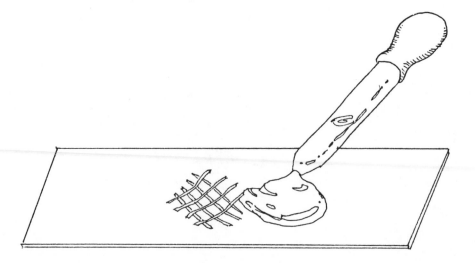

On you recording sheet, draw protists as you see them through a hand lens and microscope.

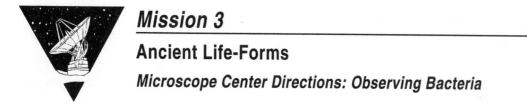

Mission 3

Ancient Life-Forms

Microscope Center Directions: Observing Bacteria

Figure 3.7—What to Look For.

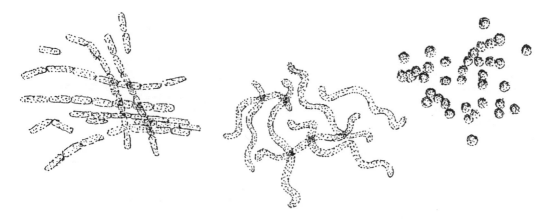

Bacteria are living organisms! Handle with care. If there is a slide already prepared, make sure the slide is still moist. If not, add one drop of water to the edge of the cover slip, as shown below.

Figure 3.8—Adding Moisture to a Slide.

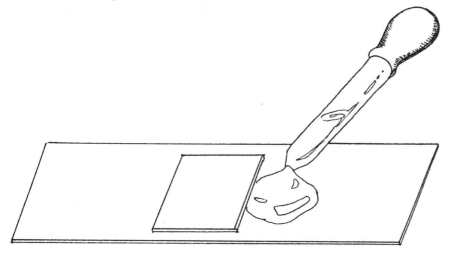

If you need to prepare a slide, put one small drop of water with bacteria on the slide and cover with a slip.

On your recording sheet, draw bacteria as you see them through a hand lens and through a microscope.

From *How Might Life Evolve on Other Worlds?* © 1995. Teacher Ideas Press. (800) 237-6124.

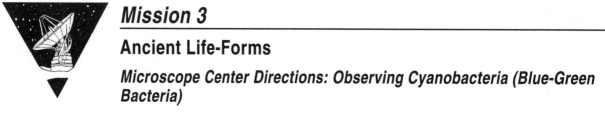

Mission 3

Ancient Life-Forms

Microscope Center Directions: Observing Cyanobacteria (Blue-Green Bacteria)

Figure 3.9—What to Look For.

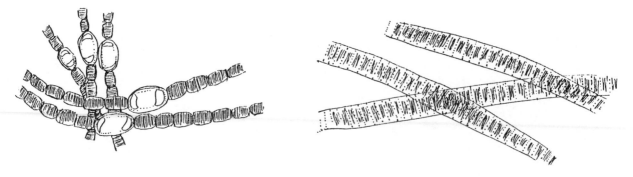

Cyanobacteria (blue-green bacteria) are living organisms! Handle with care. If there is a slide already prepared, make sure the slide is still moist. If not, add one drop of water to the edge of the cover slip, as shown below.

Figure 3.10—Adding Moisture to a Slide.

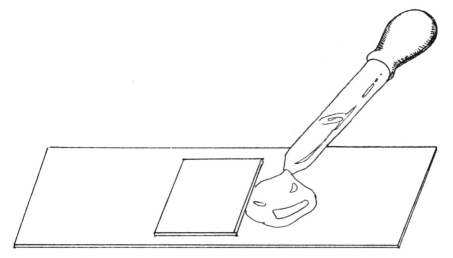

If you need to prepare a slide, and your blue-green bacteria are in *clumps*, use your medicine dropper. Aim for a small clump and draw up a droplet of the water with the blue-green bacteria in it. Cover with a slip. If your blue-green bacteria are in *long strands*, gently snip off a short piece and place it on the center of the slide. Add a small drop of water from the cyanobacteria (blue-green bacteria) container and cover with a slip.

On your recording sheet, draw cyanobacteria (blue-green bacteria) as you see them through a hand lens and through a microscope.

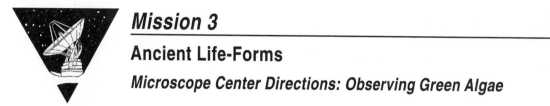

Mission 3

Ancient Life-Forms

Microscope Center Directions: Observing Green Algae

Figure 3.11—What to Look For.

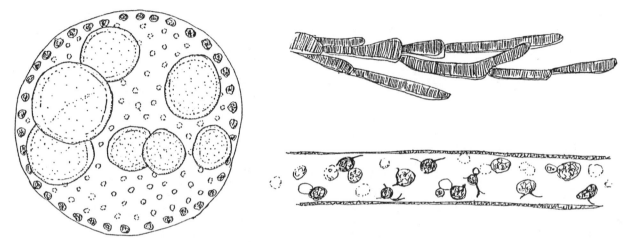

Green algae are living plants! Handle with care. If there is a slide already prepared, make sure the slide is still moist. If not, add one drop of water to the edge of the cover slip, as shown below.

Figure 3.12—Adding Moisture to a Slide.

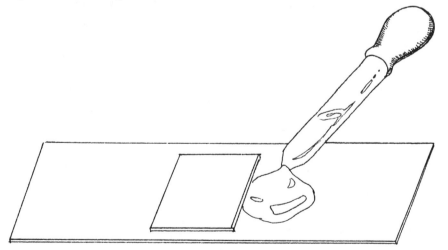

If you need to prepare a slide, and your green algae are in *clumps*, use your medicine dropper. Aim for a small clump and draw up a droplet of the water with the algae in it. Cover with a slip. If your green algae are in *long strands*, gently snip off a short piece and place it on the center of the slide. Add a small drop of water from the green algae container and cover with a slip.

On your recording sheet, draw green algae as you see them through a hand lens and through a microscope.

From *How Might Life Evolve on Other Worlds?* © 1995. Teacher Ideas Press. (800) 237-6124.

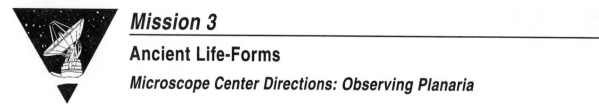

Mission 3

Ancient Life-Forms

Microscope Center Directions: Observing Planaria

Figure 3.13—What to Look For.

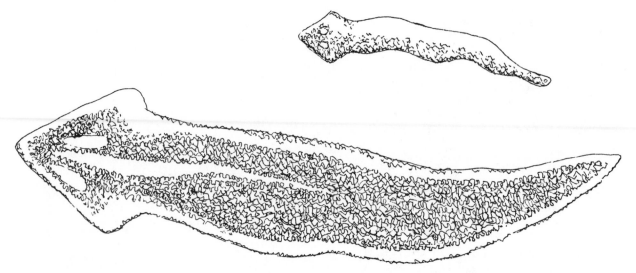

Planarians are living animals! Do not touch them! They die at the gentlest touch! If the planaria have already been fed, make sure there is water in the Petri dish. If not, add several full eyedroppers from the planaria container. Observe with a hand lens. If the planaria have not been fed, put a pinhead-sized piece of food into the Petri dish. Observe with a hand lens.

On your recording sheet, draw a view of planaria as you see them through a hand lens.

Mission 3

Ancient Life-Forms

Mission Briefing

Name:

Date:

Lorraine Olendzenski, Biologist on the SETI Academy Team.

If you were an extraterrestrial who visited Earth 730 million years ago, you would not have found any people, nor would you have found cats or dogs or monkeys or iguanas. They hadn't yet evolved! Actually, you would have found almost nothing except for tiny microscopic life-forms. The purpose of this mission is for you to imagine that you are investigating Earth 730 million years ago and record the various life-forms you discover.

What Do You Think?

1. You will be using microscopes and hand lenses to observe the descendants of ancient organisms. Below are drawings of what you might see. Which life-form do you think evolved first, second, and so on? Why do you think that?

Figure 3.14—Sample Organisms.

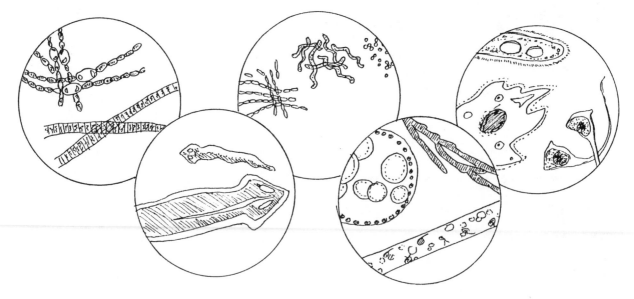

Mission 3

Ancient Life-Forms

Microscope Center Recording Sheet A

Name:

Date:

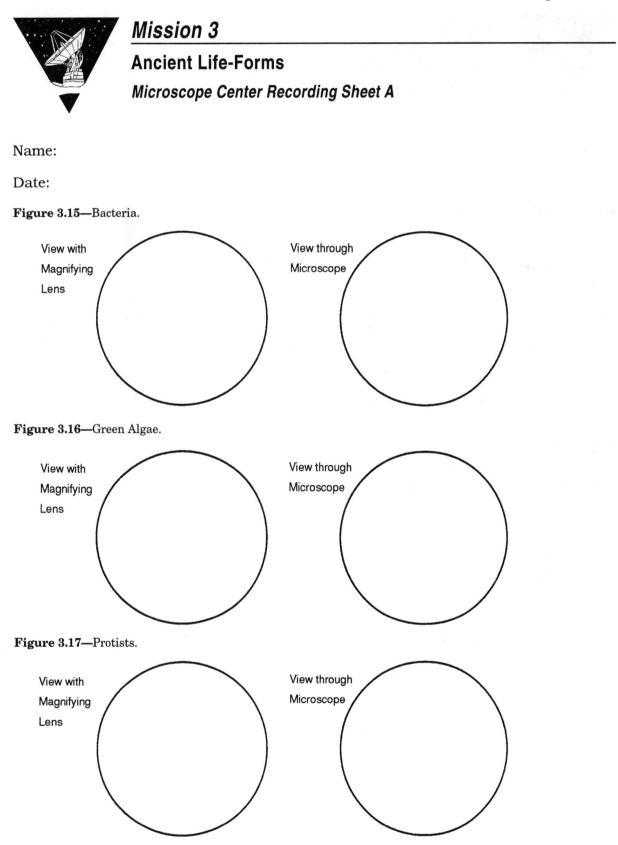

Figure 3.15—Bacteria.

View with
Magnifying
Lens

View through
Microscope

Figure 3.16—Green Algae.

View with
Magnifying
Lens

View through
Microscope

Figure 3.17—Protists.

View with
Magnifying
Lens

View through
Microscope

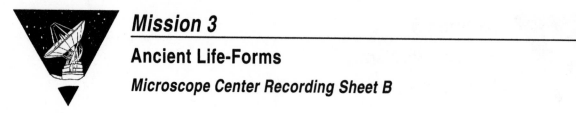

Mission 3

Ancient Life-Forms

Microscope Center Recording Sheet B

Name:

Date:

Figure 3.18—Cyanobacteria (Blue-Green Bacteria).

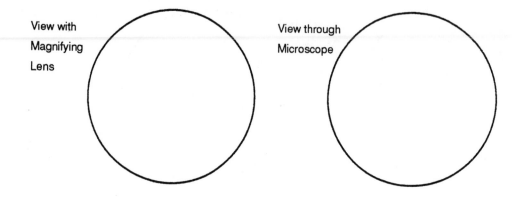

Figure 3.19—Planaria.

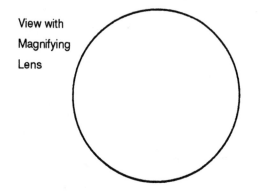

Mission 3

Ancient Life-Forms

What Do You Think, Now?

Name

Date:

After you have completed this mission, please answer the following questions:

1. How do the life-forms you observed with microscopes and hand lenses differ from each other?

2. Which life-form do you think evolved first, second, and so on? Why do you think that?

Mission 4

Who Changed Earth's Atmosphere?
Who Was the O₂ Culprit?

Overview

Notes

In mission 3, students studied microorganisms.

Science lessons about evolution usually emphasize how organisms adapt to their environment. From such lessons, we often assume that the environment shapes life. We also assume that living things cannot affect the physical environment in any significant way. However, if we look back far enough into Earth's history, we discover something quite remarkable: Early microbial life on Earth profoundly changed its environment by slowly creating an atmosphere that contained molecular oxygen.

In mission 4.1, students conduct an experiment to determine which of these microorganisms generate oxygen. They then learn which is oldest, according to the fossil records. This combination of evidence—a laboratory experiment and fossil evidence—leads students to conclude that one of the organisms is of a kind that radically changed the Earth's atmosphere and unintentionally poisoned most other life-forms! This microorganism is the "O₂ Culprit"! In mission 4.2 (four to seven days after mission 4.1), students comment on changes they see occurring in their experiment and begin to track down the "O₂ Culprit." In mission 4.3, students see an image show that describes the formation of Earth and the early stages of the evolution of life on Earth.

Concepts

- Some microscopic organisms give off oxygen.

- The oxygen in Earth's atmosphere that keeps us alive today was created by organisms like these.

- Plants still provide oxygen today.

Skills

- Conducting experiments.

- Reasoning scientifically.

- Analyzing specimens.

Mission 4.1

Materials

For the Class

- All the organisms from mission 3

- 10 Erlenmeyer flasks

- 10 clamps or clothespins

- 10 single-holed rubber stoppers to fit flasks

- 10 2-inch glass tubes to fit stoppers

- 10 1-foot sections of aquarium tubing to fit glass tubes

- 4 clean buckets

- 3 rolls of masking tape

- 5 markers

- Dechlorinated water

- Grow lamp or sunny window

For Each Student

- SETI Academy Cadet Logbook

- Pencil

Getting Ready

A Few Weeks Before Class

1. Arrange to have a volunteer to help with setup and to serve as an assistant during the experiment.

2. If obtaining equipment is difficult, contact a local junior high school or high school and arrange to borrow their equipment.

Teacher's Note: *Mission 4.1 is time-consuming (average length is 65 minutes). Schedule extra time for this experiment.*

3. Order or culture the organisms (see appendixes). Blue-green bacteria are *vital* for this
   experiment.

One or More Days Before Class

1. Prepare all the stoppers as shown in figure 4.1
   by inserting a glass tube into the hole. A drop
   of liquid soap on the tube and a twisting motion when inserting it eases this process. Be
   sure to wash off all the soap. Fit the end of the
   flexible aquarium tubing over the end of the
   glass tube. Perhaps have a parent or high-
   school-age volunteer to do this part of the
   setup.

Figure 4.1—How to Prepare a Stopper.

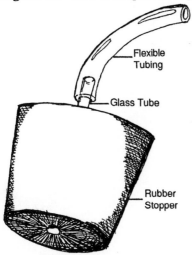

Just Before the Lesson

1. Fill the 4 buckets with dechlorinated water.
   (Remove chlorine compounds if necessary in
   your area. See Appendix.)

2. Gather all glassware onto a table or cart. In a
   demonstration area, set out 2 Erlenmeyer
   flasks, 2 prepared single-holed rubber stoppers, 2 clamps, a marker, and masking tape.

Classroom Action

1. **Mission Briefing**. Have the class refer to the
   "Mission Briefing" for mission 4 in their student logbooks while one student reads it
   aloud.

2. **What Do You Think?** Have students answer the pre-activity questions on the "Mission Briefing." Invite them to share their answers in a class discussion. Verify students' understanding of the mission 4 statement by asking them the following questions:

What kind of atmosphere did Earth have just after it was formed? *It was mostly nitrogen and carbon dioxide. There was little or no oxygen.*

Would you have been able to survive then? Why or why not? *No. People need oxygen to breathe.*

How did the atmosphere change over time? *The amount of oxygen increased.*

Did this change happen quickly? *No. It took almost two billion years.*

Where did the oxygen come from? *Certain kinds of microscopic organisms.*

What happened to the other life-forms on Earth when the amount of oxygen increased? *Most died, but some adapted to become capable of living in the presence of oxygen—an example of evolution.*

3. **Demonstration**. Tell students that they will perform an experiment to see which of the life-forms they studied in mission 3 give off oxygen; this will tell them which of these organisms is likely to be the same type of organism as the "O₂ Culprit" that caused the extinction of most of the other life-forms that were alive at that time.

Show students the supplies and demonstrate how to set up the experiment as shown in figure 4.2.

Point out the 2 Erlenmeyer flasks. Tell students that when they do the experiment, they will put *one* of the organisms into one of the flasks; they will put plain water into the second flask as a control. Tell students that, in the case of the green and cyanobacteria (blue-green bacteria), they will put as much as possible into the flask. Fill each flask 1/2 full with dechlorinated water. Tell students that this is the point at which they will add their organism to the flask. Next, insert the stopper into the flask and clamp the tubing. Do this to the second flask.

Figure 4.2—Setting Up the Experiment.

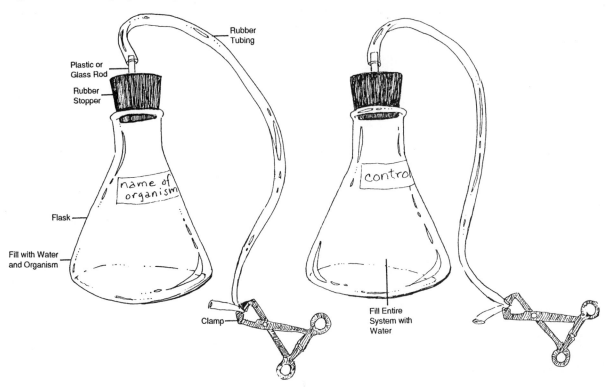

Explain that the purpose of the control flask is to be sure that any change seen in the experimental flask is caused by the organism, and not by the water. So, the same water source and the same equipment will be used, and the flasks will be put in the same place for the same length of time. Therefore, any difference between the two flasks must be caused by the organism.

Demonstrate how to use masking tape and a marker to label one flask with the name of the organism, and label the other (with plain water) "control flask."

Check for comprehension about how to set up the experiment. Be sure students understand that they need to include their assigned organism in the "experimental" flask. Tell students that they will be looking for bubbles of gas that may collect at the top of the flask or in the tubing.

Assign students to one of five teams: bacteria, cyanobacteria (blue-green bacteria), green algae, protists, and planaria. Each team sets up a gas collection experimental flask for one of the organisms and a control flask, using the technique demonstrated.

4. **Activity**. Have students do the lab. Put the experimental equipment on a window sill or under a grow lamp for four to seven days. Protect the experiment from curious students or visitors who might be tempted to pick up a flask or tamper with a clamp.

Mission 4.2

Materials

For the Class

- 6 wooden splints

For Each Student:

- SETI Academy Cadet Logbook

- Pencil

Getting Ready

Just Before the Lesson

1. Have the other identical sets of glassware and lab supplies available for the students along with masking tape. Place the buckets of water in different places around the room.

Classroom Action

1. **Discussion**. Students begin to track down the "O₂ Culprit." Set out the experimental flasks on a table where students will be able to see all of them. Have students comment on the changes they can see. In the case of the green algae and cyanobacteria (blue-green bacteria), it may be possible to see bubbles trapped on the growing organisms.

2. **Demonstration**. Show students how to test the gas in each flask to see if it is oxygen:

   a. Tap the flask gently to release any gas bubbles that may have been trapped.
   b. Have a student carefully remove the clamp from the aquarium tube, keeping the tube tightly crimped. It is essential not to release any gas at this point.
   c. Have a student light a wooden splint and blow it out, leaving it still glowing.

Teacher's Note: Remind students of precautions needed when handling flame and gas elements.

    d. Quickly straighten the crimp in the aquarium tube, directing the open end of the tube toward the glowing splint. If the gas from the tube is oxygen, the glowing splint will burst back into flame for an instant.

3. **Activity**. Have students test each flask for oxygen with splints to discover which of the organisms gives off oxygen.

4. **Lecture**. Students should find that cyanobacteria (blue-green bacteria) and green algae both give off oxygen, so they are the main suspects. Ask students how they can tell which one is the same kind of organism as the "O_2 Culprit." To find this out, students will need to know which one evolved first. Tell students that they will be watching an image show, presenting the best theory that scientists have come up with so far, based on fossil evidence.

Mission 4.3

Materials

For the Class

- *History of Earth* video

- VCR and monitor

- (optional) Transparencies of the black-line masters

- (optional) Overhead projector

- (optional) "Life Story of the Earth" video script

For Each Student

- SETI Academy Cadet Logbook

- Pencil

Getting Ready

One or More Days Before Class

1. (optional) Make transparencies of the black-line masters.

Just Before the Lesson

1. Set up the VCR and monitor. Start the *History of Earth* video at "Mission 1." Have the "Life Story of the Earth" video script handy.

2. (optional) Set up the overhead projector.

Classroom Action

1. **Video or Black-Line Masters**. Introduce the *History of Earth* video of still images. The "Life Story of the Earth" video script follows this lesson plan.

2. **Discussion**. Discuss any portions of the video that may have confused students. Use transparencies of the images as desired.

Closure

1. **Discussion**. After the image show, invite students to share their impressions and ideas. What were their initial ideas about how life formed in the first place? How have those ideas changed? Students should have an opportunity to express religious as well as scientific ideas. Invite discussion, and remind students that the fossil evidence tells us only *what* organisms formed and *when* they were abundant. Scientists still do not fully understand *how* they came about or *why* major advances in evolution occurred at specific times in the past.

2. **What Do You Think, Now?** Have students answer the post-activity questions on the logbook sheet "What Do You Think, Now?" Invite students to share their responses. Ask students how their opinions have been changed by this mission.

Going Further

Activity: The "O$_2$ Culprit"

As an art activity, have students create a "wanted" poster for the organism they found to be guilty. Have students write an autobiography of the "O$_2$ Culprit."

Mission 4

Script for Video Images

Life Story of the Earth"

Figure 4.3—About four billion years ago, Earth's surface and oceans had formed. Volcanoes belched huge clouds of carbon dioxide and sulfurous gases, forming the early atmosphere.

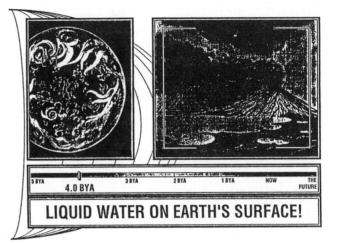

LIQUID WATER ON EARTH'S SURFACE!

Figure 4.4—As time went by, large islands started forming. By around 3.8 billion years ago, the first microscopic bacterial-like forms could be found in Earth's oceans. Nobody knows how life was created. One theory is that the first life-forms were created by chance combinations of complex molecules in tide pools or shallow seas, warmed by the Sun and energized by lightning. Another theory is that the first life was made by chance combination of complex molecules at places where hot volcanic vents heated the ocean on the sea floor.

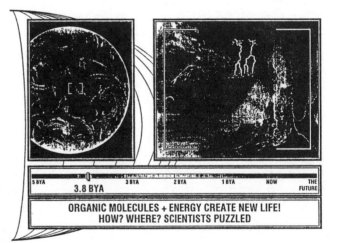

**ORGANIC MOLECULES + ENERGY CREATE NEW LIFE!
HOW? WHERE? SCIENTISTS PUZZLED**

Figure 4.5—Fossil remains of layered mounds about 1 meter in size have been discovered that are 3.5 billion years old. These mounds are called stromatolites. They are the remains of communities of millions of cyanobacteria (blue-green bacteria) that grew in layers, leaving behind stromatolites that are among the oldest evidence of life on Earth. These primitive organisms had simple cells without a nucleus.

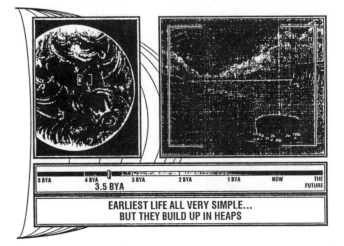

**EARLIEST LIFE ALL VERY SIMPLE...
BUT THEY BUILD UP IN HEAPS**

Figure 4.6—By about 2.5 billion years ago, the atmosphere had changed. How do we know? Large deposits of rust—iron combined with oxygen—shows that there were great amounts of oxygen in the air for the first time. The oxygen in the atmosphere must have poisoned many other life-forms at the time. At this time, huge stromatolites still existed. Eventually, slightly more complex life-forms evolved. These single cells had a nucleus. They were very much like the green algae and protists that exist on Earth today.

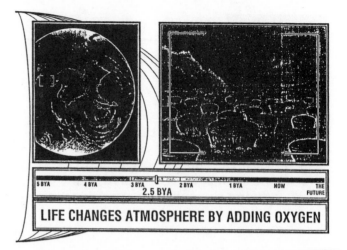

Figure 4.7—Until about 1.7 billion years ago, cells reproduced when one cell divided into two identical cells. This is shown in the top diagram. Then, certain cells evolved that reproduced in a different way. As shown in the bottom diagram, two cells combined to form a new cell that was like each of its parents, but a little different. This is called *sexual reproduction*. This kind of reproduction made it possible for millions of new species to form—and evolution really took off!

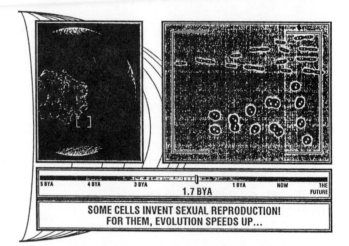

Figure 4.8—Sometimes a cell would reproduce, but would fail to divide into two separate individuals. Some of these clumps of cells had advantages over other life-forms. Eventually, multicelled organisms made of many different kinds of cells formed. The different cells specialized, forming skin, sense organs, and other body parts. By one billion years ago, the complexity of the organisms we find preserved in the fossil record increased dramatically.

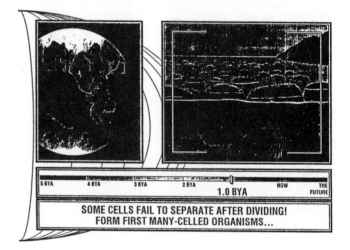

Figure 4.9—About 730 million years ago, some organisms developed the first specialized cells that worked together to form simple nerves and muscles. Nerves and muscles allowed these organisms to move about. A flatworm, much like the planaria that you saw in class, was one of these first mobile creatures.

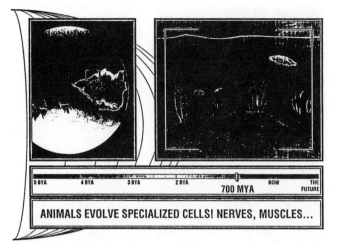

Mission 4

Who Changed Earth's Atmosphere?
Mission Briefing

Name:

Date:

Jon Jenkins, Atmospheric Scientist on the SETI Academy Team.

The first life-forms on Earth evolved in an atmosphere that would have been deadly to us. Those single-celled life-forms breathed "air" that was almost entirely composed of carbon dioxide gas. There was no oxygen at all. Then, one form of life that used photosynthesis as an energy-producing mechanism happened to evolve. This life-form used carbon dioxide as raw material and produced oxygen as a waste product. At first, very little happened to other life-forms, because the oxygen combined with iron that was dissolved in the primitive oceans. This created rust, which settled to the bottom of the ocean. Eventually, all the iron in the water was used up and the oxygen started building up in the atmosphere. This process took almost two billion years. Unfortunately for the other life-forms that were around then, oxygen was toxic! You could say that the first life-form to give off oxygen (O_2) unthinkingly poisoned most of the other forms of life that had evolved up to that time. The challenge of this next task is for you to determine who that "O_2 Culprit" was.

What Do You Think?

1. Which early life-form do you suspect was the "O_2 Culprit"?

2. How do you know?

Mission 4

Who Changed Earth's Atmosphere?

What Do You Think, Now?

Name:

Date:

After you have completed this mission, please answer the following questions:

1. Which early life-form was the "O_2 Culprit"?

2. How do you know?

Mission 5

Fossils! Layers of Ancient Life Buried in the Earth

Overview

But what *is* a fossil? How are fossils made? How do we know how old they are? In mission 5, students get an introduction to fossils and the fossil record. In mission 5.1, they make a "Fossil Jar" showing how sedimentary rock layers form with fossils inside. In mission 5.2, they make a "Fossil Layer Cake" showing how rock layers can change their shape and location after they are formed.

Notes

In mission 4, students conducted an experiment to determine which organism generated oxygen and helped to form the atmosphere that we have today. They used the fossil record to help find the O_2 culprit.

Concepts

* Living organisms (microorganisms, plants, and animals) can become fossilized after they die.

* Sedimentary rock forms as sediment that builds up and eventually turns into stone because of heat and pressure.

* Living organisms that are buried in a layer of sediment may fossilize inside the rock as it forms.

* Sedimentary rock forms in layers, with the younger layers forming on top of the older layers.

* It takes a *long* time (millions of years) for enough sediment to build up to turn older, deeper layers into stone.

* Each layer of sedimentary rock, with its fossils, is like a photograph of one period of time in Earth's history.

* After sedimentary rock layers form, the movements of Earth may change the shape and location of the layers.

Skills

- Doing a simulation.

- Recording data into a table.

Mission 5.1

Materials

For the Class

- Clock

- (optional) Real fossils

- (optional) Small pieces of shale, small pieces of sandstone

- (optional) Transparency of "Making a Fossil Jar" logbook sheet

- (optional) Overhead projector

For Each Pair or Group of Students

- Mason jar or other glass or clear-plastic jars

- Sand

- Dirt

- Tiny plastic animals (or shells); at least three kinds

- Water

For Each Student

- SETI Academy Cadet Logbook

- Pencil

Getting Ready

One or More Days Before Class

1. Test the sand and dirt to be sure that they form distinct layers. The dirt should be moist (dry peat moss or planting mix may float). Test the plastic animals; they must sink.

Classroom Action

1. **Mission Briefing**. Have the class refer to the "Mission Briefing" for mission 5 in their student logbooks while one student reads it aloud.

2. **What Do You Think?** Have students answer the pre-activity questions on the "Mission Briefing." Invite them to share their answers in a class discussion.

3. **Discussion**. Ask students to imagine that they are paleontologists. Paleontologists are fossil hunters! Draw a simplified version of figure 5.1 on the chalkboard. They have just found the fossil remains of two animals. One was 3 feet below the surface and one was 10 feet below the surface. Ask students which fossil is older and why they think so. Tell students that determining this is not as simple as they might first think! Ask them what they think is more important: how deep each fossil is buried or which fossil is on top of the other.

Figure 5.1—Fossil Layers.

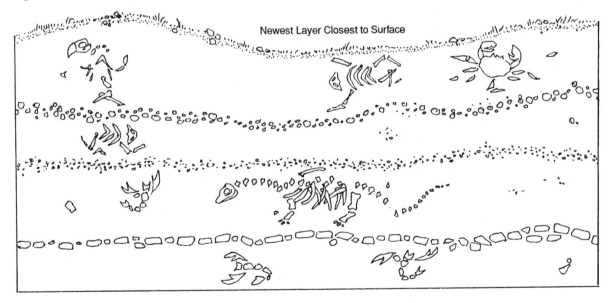

4. **Lecture**. How does a fossil form? Tell students that when a plant or animal dies, it may be gradually covered with leaves, mud, dust, and even volcanic ash. Organisms that die underwater may be covered by sediment. The soft parts will rot, but the hard parts sometimes remain. Under certain conditions, these buried organisms may fossilize. They may turn to stone or create an impression, or the bones and teeth

themselves may simply remain as bones and teeth and not turn to stone. If possible, show students real fossils.

The original sediment may turn into rocks due to pressure and heat; sand can become sandstone, and mud can become shale. If possible, pass around samples of sandstone and shale for students to see and touch.

Over the years, more and more material covers the fossil, so that the older plant and animal fossils are deeper in the ground than the more recent ones. For instance, in figure 5.1, the dinosaur fossils are found in the same layer as small mammals, but they are both deeper than some large mammals that evolved more recently, including people. However, in some areas, like the Badlands in South Dakota, the dinosaur fossils are exposed on the surface of the ground or are only shallowly buried near the surface. How can this be? Tell students that they will investigate this in the next two activities.

5. **Demonstration**. As a demonstration, make a large Fossil Jar as shown in figure 5.2. Use the clock for timing. If only doing the demonstration have students record the information on their logbook sheets (see page 85). For each sediment layer, record on the chalkboard the time at which pouring begins and the time at which pouring ends. Begin by filling a glass jar about 1/2 full of water. Then pour about 1/2 cup of sand in, very slowly, so that the individual grains can be seen falling to the bottom. Explain that this could represent a marsh, with a little stream bringing in tiny particles called *sediment*. Let a layer build up until it is distinctly visible.

Now place one or more small plastic animals or shells of the same kind on the surface of the sediment layer. This represents an organism (microorganism, plant, or animal) that has died. These objects may be placed against the glass so that they will remain visible after they are buried. Tell students that when they build their own Fossil Jars they don't have to put an organism on each layer, and on some layers they may want to put more than one organism. After all, it was chance that any given organism became fossilized. Continue to add sand until the animals or shells are covered.

Figure 5.2—Fossil Jar.

Next, begin pouring dirt in slowly until a layer forms that is distinctly visible. Add animals or shells different than the ones in the first layer of sediment. These represent a different organism. Cover these with more dirt. Repeat this process several times, depending upon the jar size and the number of kinds of animals and shells.

When students figure out the age of their fossils, the unit of time will be "minutes ago," a direct measure of how long it took these layers to form. This will show students why each geological period has a beginning date and an ending date. Demonstrate how to determine the age of the fossils in each sediment layer by subtracting pouring finish times from the current time.

(optional) If students are not familiar with filling in data tables, use a transparency of "Making a Fossil Jar" (page 84) and an overhead projector to demonstrate the technique.

(optional) It may be easier for younger students to "round-off" seconds to the nearest minute to make for easier subtraction when calculating the age of their fossils.

6. **Activity**. Have students follow the directions on "Making a Fossil Jar" logbook sheet (see page 84) and create their own Fossil Jars. Letting students take home their Fossil Jars will help avoid a cleanup problem!

7. **Discussion**. Ask students what could happen to real animals that were buried under sediments like the plastic animals in the jars. If possible, show students real fossils. Ask students what might happen to the sand. To help them understand how sand could become sandstone, rub a piece of sandstone, if available, until sand grains come off. Ask students what might happen to the mud. Show students a piece of shale, if available. Explain that it takes time, heat, and pressure to turn sand and mud into rock. Ask students how long they think it might take for sediment to build up enough to cover an animal and then to turn it into stone. *Millions of years!*

Explain that one way paleontologists tell how one of two fossils is older is to notice which one is buried in a lower layer of sedimentary rock. Ask students if they think that Earth always stay the same, or if they think that sometimes it moves. Ask students what they think an earthquake would do to fossils. Tell students that, in the next activity, they will see how moving earth on Planet Earth can change the location of fossils.

Mission 5.2

Materials

For Each Pair or Group of Students

- 6 sheets (or more) each of red, yellow, blue, and black construction paper

- Scissors or a paper cutter

- Tape

- Markers or crayons: red, yellow, and blue

For Each Student

- SETI Academy Cadet Logbook

- Pencil

Getting Ready

No preparation is necessary.

Classroom Action

1. **Lecture**. Tell students that scientists often find fossils of seashells and marine fish in the rocks on mountaintops, including the Andes Mountains. Ask them how fossils that formed under the sea could end up on a mountaintop. Remind students about earthquakes.

    If the class has a background in plate tectonics (see mission 11 in *The Evolution of a Planetary System*, the first book of the *SETI Academy Planet Project*), build upon that. If not, briefly explain the basics: The crust of Earth is broken into pieces called *tectonic plates*; subduction, earthquakes, volcanoes, and mountain-building processes occur at plate boundaries. The major point to stress is that fossil-bearing sedimentary rock layers can move around after they are formed. What was once the sea floor can be pushed up to form a mountain. Erosion can strip away more recently formed layers of rock, leaving older rock at the surface. Geologists are aware of many such changing forces.

2. **Activity**. Give students at least 6 sheets each of red, yellow, and blue construction paper (more would be better). Each color will represent a different type of sedimentary rock. Have students cut out the pictures of fossils ("Cutouts" logbook sheet); they do not need to be precise in their cutting. They should tape one kind of fossil on one color of paper. Each color paper should only have one kind of fossil: red = clams; yellow = trilobites; blue = shark teeth. Have students follow instructions on page 88 of their logbooks.

    Have students stack the layers: Red is the bottom, yellow in the middle, and blue on top. This represents the original formation of flat layers of sedimentary rock. Have them follow the directions in the student logbook, simulating different movements of Earth. Demonstrate these, and other more complex movements, if it seems that students need help.

To conserve paper, have students tape the fossils onto the papers. The taped-on fossil pictures can be removed later, making the paper available for reuse. When layering the papers, insert some layers without fossils on them, making that paper available for reuse.

3. **Discussion**. Ask students what the three colors of paper represent. Ask students which layer is the oldest, which layer is the youngest, and which layer is middle-aged; and why. *The red, yellow, and blue papers all represented layers of sedimentary rock. The red is the oldest because it formed first and was on the bottom. The blue is the youngest because it formed last and was on the top. The yellow was the middle-aged because it was in the middle and formed after the red and before the blue.*

Ask students if any of the changes they made, such as the mountain and the canyon, changed the *order* of the layers. *No. It was always red-yellow-blue.* Were the shark teeth ever found under the trilobites? *No. They were always on top.* Did the order of the fossils themselves ever change? *No.* Ask students which fossils are the oldest and how they can tell. *The clams because they were always on the bottom, in the rock that formed first.*

Ask students if any of the changes they made, such as the mountain and the canyon, changed *where* fossils would be found by a paleontologist. *Yes. Some fossils ended up on top of the mountain, or in the canyon. The paleontologist might find shark teeth on the plain, on the mountaintop, or in the canyon.*

Ask students if any of the changes they made, such as the mountain and the canyon, changed how *deep* a fossil would be found. *Yes. After the volcano, everything was buried deeper in the Earth. Also, after the erosion, the trilobites were right at the surface!*

4. **Lecture**. Explain that when paleontologists tell how one of two fossils is older by noticing which one is buried in the lower layer of rock, they are getting a *relative* date. This does not tell them how long ago either rock layer formed. In recent years, scientists have been able to determine the actual age of many fossil

Teacher's Note: Dating by radioactive decay is a fairly complex subject for children to understand, but they should be aware that scientists do have methods for determining absolute dates for rocks and the fossils they contain.

plants and animals by measuring the amounts of certain radioactive materials left within certain kinds of rock (e.g., igneous rock, from lava flows) above or below the sedimentary rock that the fossils are found within. This is how we know, for example, that the dinosaurs died out about 65 million years ago.

Closure

1. **Review**. Have students explain why, in some areas, like the Badlands in South Dakota, the dinosaur fossils are exposed on the surface of the ground or are only shallowly buried near the surface. Ask them if they can tell how old a fossil is by how deep it is buried.

2. **What Do You Think, Now?** Have students answer the post-activity questions on the logbook sheet "What Do You Think, Now?" Invite students to share their responses. Ask students how their opinions have been changed by this mission.

3. **Preview**. Tell students that, in the next mission, they will discover why animals change over time and thus why different animals appear in different rock layers in the fossil record. They will become birds and play a game to simulate natural selection!

Going Further

Activity: Excavating Fossils in a Jar

Have one class give their Fossil Jars to another class that has not had this lesson and challenge them to uncover the fossils and reconstruct the order in which they were buried.

Research: Fossils on Mars!

The conditions on Mars today make the existence of Earth-type life unlikely. However, the conditions on Mars in the past were more like those on Earth. Some scientists believe that life may have begun on Mars and then become extinct. If so, there might be fossils on Mars. A good place to look would be the bottoms of the ancient riverbeds. Have students research and report to the class on articles and books about this subject.

Research: The Grand Canyon

In the Grand Canyon of Arizona, many layers of rock are cut through by the Colorado River. Students can see this in any picture of the Grand Canyon. They can discover how long ago each layer formed, and what kinds of fossils appear in each layer. Have students research and report to the class on articles and books about this subject.

Field Trip: Real Fossils!

If there is a museum in the local area, it may well have real fossils—even mounted dinosaur skeletons. Take students on a field trip. In some parts of the country, it is possible to take students on a real fossil hunt for simple abundant fossils such as shark teeth or clams. Contact local nature centers.

Activity: Fossil Hunt in a Box

Use chicken bones, shells, and other objects to represent fossils. Create a shoe box with layers of dry dirt and sand with embedded fossils. Teachers may make these boxes and then have students excavate them to see which fossils are on top of others, or have teams of students construct boxes and trade with other teams before excavation. The challenge is to note which layer each fossil is found in and then deduce relative ages (assuming no major geological shifts).

For an extra challenge, embed the fossils in wet potter's clay and let it dry before excavation! Students will then need tools to dig out the fossils. If the fossils are fragile, they can actually use brushes to remove the fossils from the rocks, as paleontologists often have to do.

Mission 5

Fossils!

Mission Briefing

Name:

Date:

Dr. Rufus Catchings, Paleontologist on the SETI Academy Team.

During your last mission, you were briefed on the fossil record. This is the history of the evolution of life on Earth that is "written in the rocks." Scientists use this record to see when different kinds of organisms first appeared on Earth, and to figure out their evolutionary ancestry. When we explore other planets, we might find fossil records that show how evolution proceeded on those planets. In this mission, your task is to discover how fossils are made and how scientists can tell how old they are.

What Do You Think?

1. Can you tell how old a fossil is by how deep in the ground it is buried?

2. If one fossil is found on top of another fossil, which one is older?

Mission 5

Fossils!

Making a Fossil Jar

1. Fill a glass jar about 1/2 full of water.

2. Pour sand in very slowly, so that the individual grains can be seen falling to the bottom. Let a layer build up until it is distinctly visible.

3. Place one or more of the *same* kind of small plastic animal or shell on the surface of the sand layer. The animals may be placed against the glass so they will remain visible from the side after they are buried.

4. Write down the name of the animal in the second table on your "Making a Fossil Jar" logbook sheet.

5. Continue to add sand until the animals are covered.

6. Write down what time it is now in the first table on your logbook sheet.

7. Repeat steps 2 to 6 using dirt instead of sand. Then do it one more time with sand. Be sure to record all your data!

8. For fun, you may add even more layers of dirt and sand, if you have a big jar and more kinds of animals.

9. Congratulations! You have created a Fossil Jar!

Mission 5

Fossils!

Making a Fossil Jar

Name:

Date:

1. Record your data for a "Time Table."

Table 5.1—Time Table Data Record.

What You are Doing	Time When Finished
Pouring first layer	
Pouring second layer	
Pouring third layer	

2. Record your data for a "Fossil Table."

Table 5.2—Fossil Table Data Record.

Sediment Layer	Name of Fossilized Animal Found
First Layer (sandstone)	
Second Layer (shale)	
Third Layer (sandstone)	

3. Now figure out how old each fossil is in "minutes ago" by subtracting each time measurement from the current time.

   Current time: _____

Table 5.3—Current Time.

Name of Fossilized Animal	Age (in "minutes ago")

Mission 5

Fossils!

Cutouts

Figure 5.3—Fossil Illustration Sheet.

From *How Might Life Evolve on Other Worlds?* © 1995. Teacher Ideas Press. (800) 237-6124.

Mission 5

Fossils!

Making a Fossil Layer Cake

1. Cut out the pictures of fossils.

2. Tape all the clam pictures to red construction paper. You may put one or more of these clams on each red page. Some red pages may be left blank.

3. Tape all the trilobite pictures to yellow construction paper.

4. Tape all the shark tooth pictures to blue construction paper.

5. Make a big stack using all the papers; all group members should put their papers into one stack. Put the red papers on the bottom, one at a time. Add the yellow papers, one at a time. Finish with the blue papers on top, one at a time.

6. Draw a side view of your model Fossil Layer Cake on your "Making a Fossil Layer Cake" logbook sheet.

7. Make a mountain by bending your Fossil Layer Cake upward in the middle, so it looks like an upside-down letter *U*.

8. Draw a side view of your mountain on your logbook sheet.

9. Flatten the Fossil Layer Cake into a plain again.

10. Make a canyon by bending your Fossil Layer Cake downward in the middle and flattening the edges, so it looks like a letter *U*.

11. Draw a side view of your canyon on your logbook sheet.

12. Flatten the Fossil Layer Cake into a plain again.

13. A volcano covers your rocks with lava! Add several sheets of black construction paper on top of the stack.

14. Draw a side view of your Fossil Layer Cake after this volcanic eruption on your logbook sheet.

15. Time passes, and rain and wind wear away the lava and the upper layer of sedimentary rock. Remove the black papers and the blue papers.

16. Draw a side view of your Fossil Layer Cake after this erosion.

Mission 5

Fossils!

Making a Fossil Layer Cake

Name:

Date:

1. Draw a side view of what your Fossil Layer Cake model looks like just after you make it. Include a picture of the kind of fossil found in each layer. What kind of fossils are found near the surface (top of the stack)?

2. Draw what your model looks like after you make a mountain. Include a picture of the kind of fossil found in each layer. Are the fossils found on the mountaintop the same as were found on the plain above?

3. Draw what your model looks like after you make a canyon. Include a picture of the kind of fossil found in each layer. Are the fossils found on the floor of the canyon the same as were found on the plain?

4. Draw a side view of what your model looks like after the volcanic eruption. Include a picture of the kind of fossil found in each layer. What kind of fossils are found at the surface (top of the stack)?

5. Draw what your Fossil Layer Cake looks like after the erosion. Include a picture of the kind of fossil found in each layer. What kind of fossils are found near the surface (top of the stack) now?

6. Did anything change the *order* of the layers that have fossils?

7. Did anything change *where* fossils would be found?

8. Did anything change how *deep* fossils would be found?

From *How Might Life Evolve on Other Worlds?* © 1995. Teacher Ideas Press. (800) 237-6124.

Mission 5

Fossils!

What Do You Think, Now?

Name:

Date:

After you have completed this mission, please answer the following questions:

1. In the Fossil Jar, what did the sand represent? What *could* happen to the sand if this simulation were real?

2. What did the dirt represent? What *could* happen to the dirt if this simulation were real?

3. If all dinosaurs were extinct by 65 million years ago, why are there dinosaur bones near the surface in the Badlands of South Dakota?

4. Can you tell how old a fossil is by how deep it is buried? If not, how *can* you tell?

Mission 6

Natural Selection
Who Will Survive?

Overview

Students have not been given a mechanism for evolutionary change. The two main theses of Charles Darwin's *Origin of Species* are that organisms are products of a history of *descent with modification* from common ancestors, and that the main mechanism of evolution is the *natural selection* of hereditary variation. In mission 6, the students will be introduced to natural selection as a way in which organisms can change over time. In missions 8 and 11, they will explore descent with modification.

Concepts

- One *kind* of any microorganism, plant, or animal is a species.

- Individuals within any species look and act differently from each other, just as individual people within our species look and act differently.

- Some individual differences are more necessary than others for successful survival in the current environment.

- Natural selection means that the *environment* selects the differences that allow the best chance of survival.

- Individual differences are passed on from successful parents to children. In general, children resemble their parents.

Skills

- Recording data into a table.

- Forming hypotheses.

- Deductive reasoning.

In missions 1-5, students saw that the fossil record shows that the first life-forms were microscopic and single-celled, and that some later life-forms became multicellular, bigger, and more complex plants and animals.

Mission 6

Materials

For the Class

- Extra chopsticks, pliers, salad tongs, beans, peas, and paper cups to be used as needed

For Each Group of Students

- Pair of chopsticks

- Pair of pliers

- Salad tongs

- 50 each of three different kinds of dried seeds: brown beans, white beans, and green peas

- 3 paper or plastic cups

- Carpet scrap or piece of cloth about 2 feet square; 1 piece each of brown, white, and green carpet or cloth will work best (each group gets one color)

- Stopwatch or clock with a second hand

- Large plastic bag

- (optional) Calculator

- (optional) Other grasping tools such as tweezers

For Each Student

- SETI Academy Cadet Logbook

- Pencil

Getting Ready

Before the Lesson

1. Divide the class into groups of six students each.

Classroom Action

1. **Lecture.** What is evolution? Ask students for their ideas about what might cause one kind of animal to evolve into a different kind of animal over millions of years. Give them time

to share their ideas. Introduce the idea that small changes may add up to big changes, given a lot of time.

(optional) Emphasize the distinction between a theory and a hypothesis. A *theory* is a broad scientific concept about nature that has been supported by a large amount of evidence and has not been contradicted by any evidence. *Evolution* is a theory. This means that it is accepted by scientists as the best possible explanation for the evidence they can observe. It has been supported by a large amount of evidence, and has not been contradicted by any evidence. A *hypothesis* is an educated guess about some particular occurrence in nature. Theories may be useful for hundreds of years before being modified or replaced, but hypotheses change from day to day as scientists conduct new experiments or make additional observations.

2. **Mission Briefing**. Have the class refer to the "Mission Briefing" for mission 6 of their student logbooks while one student reads it aloud.

3. **What Do You Think?** Have students answer the pre-activity questions on the "Mission Briefing." Invite them to share their answers in a class discussion.

4. **Lecture**. Tell students about the process called *natural selection*: Animals that are less well adapted to the current environment tend to die out; those that are more well adapted tend to survive and pass on those qualities that enabled them to survive to their offspring. Natural selection is a factor in the theory of evolution, which is the best scientific explanation for how all of today's microorganisms, plants, and animals developed from one-celled organisms like those studied in previous missions. According to this theory, in just a few years, natural selection can bring about small changes in species. Over millions of years, natural selection can result in entirely new species.

5. **Demonstration**. This laboratory helps students understand the process of natural selection and how it can lead to genetic changes in a population. In this activity, students become

seed-eating finches with three different beak types in their species: one represented by chopsticks, one by salad tongs, and one by pliers. One of these should prove to be the best at picking up seeds (dried beans and peas). Demonstrate the entire Hungry Finch Game, or at least demonstrate "proper" and "illegal" ways of picking up seeds. The seeds represent three variations of one plant species. They have different sizes and colors. These differences may help particular seeds to survive on particular colors of carpet or cloth. The white, brown, and green carpets or cloths represent three different environments.

If necessary, show students how to record data in table form. Show how to count the seeds gathered in the cups as "prey" and the seeds left on the carpets or cloths as "survivors."

Play one "trial run" of the Hungry Finch Game to be sure that everyone understands the complex instructions.

6. **Activity**. Distribute the carpets or cloths equally to different teams (one per team). Have students follow the directions and play the Hungry Finch Game.

7. **Discussion**. Because there are likely to be differences as well as similarities in the results of the Hungry Finch Game, it is important to share class data and experiences. Note that evolution does not always proceed in exactly the same way. Because of the excitement of the game, students may not notice that differences in the environment, such as the color of the carpet, provide a degree of camouflage to particular seeds.

Ask students how natural selection changed the *finches*. The finches started with an equal number of individuals of each beak difference. Which differences become more common in the finches? Why? Which differences become less common in the finches or were entirely eliminated? Why? Which differences remained about the same in the species? Why? Are the finches *evolving*?

Ask students how natural selection changed the *plant species*, represented by the seeds. The seeds started with an equal number of individuals of each size and color difference. Which differences become more common in the

Teacher's Note: There are two forms of the data sheet. One lists the recommended "grabbers" and seeds. The other is blank. Use the blank form if you are substituting other "grabbers" or seeds.

Teacher's Note: After 20 seconds, stop all feeding. Students should have picked up about half of the seeds; if they have picked up more than 60 percent or less than 40 percent, then adjust the time interval and start over. It is important to adjust the time interval to allow for varying motor skills among students in different age groups. Do a trial run with the whole class to establish this time interval, and then have all the groups play the game. For particularly aggressive game players, play the game on the floor to ensure that students have equal access to the carpets or cloths.

seeds? Why? Which differences become less common in the seeds or were entirely eliminated? Why? Which differences remained about the same in the species? Why? Are these seeds (plants) *evolving*?

Ask students if a different color of carpet (environment) changes the results. Compare the results of each group to find out.

8. **Lecture**. Relate the Hungry Finch Game to a real situation. Explain what kinds of environmental differences might be important to these real organisms' survival and reproduction. A few years ago, naturalist Peter Grant observed finches on an island over a period of years. Then a dry spell killed all the plants except for those with extra-large, drought-resistant seeds. Because finches eat seeds, those with small beaks died. Those that happened to have larger beaks survived because they could crack the large seeds. The offspring of these finches tended to have large beaks too, and so, after a couple of years, only finches with larger beaks could be found. The small-beaked finches had died out.

Closure

1. **Review**. Relate the small changes that students saw in the Hungry Finch Game to the big changes that create new species. Remind them that little changes can add up to big changes over a long period of time: 1 + 1 + 1 + 1 + 1 + ... will eventually reach one million!

2. **What Do You Think, Now?** Have students answer the post-activity questions on the logbook sheet "What Do You Think, Now?" Invite students to share their responses. Ask students how their opinions have been changed by this mission.

3. **Preview**. Tell students that, in the next mission, they will see, as they continue their exploration of the fossil record, the results of natural selection, and the origin of many new species over time.

Going Further

Research: "DDT Doesn't Bug Me!" Bugs

Before the pesticide DDT was banned, it was becoming less and less useful, as more and more insects became resistant to it. Challenge students to see how this is an example of natural selection leading to evolutionary change.

Activity: Hungry Extraterrestrials

Have students play a version of the Hungry Finch Game and call it the Hungry Extraterrestrial Game. Student groups each create an extraterrestrial creature and devise an adaptation that would enable the extraterrestrial organism to survive in a changing environment. Students choose "tools" to serve as the adaptation, decide what type of food their extraterrestrial organism might eat, and concoct some kind of worldly natural disaster that their extraterrestrial may or may not survive, such as drought, flood, or change in climate. When each group has discussed what the extraterrestrial habitat looks like, they are ready to play the Hungry Extraterrestrial Game.

Activity: Peppered Moths

The English peppered moth was commonly white, because that color camouflaged it on light tree bark, and predatory birds ignored white moths. In a famous example of natural selection, as the Industrial Revolution blackened tree trunks with soot, the formerly rare black form of the peppered moth now had the advantage and became more numerous. Students can cut out black and white moths from paper and put them on black or white construction-paper "tree bark." Then, one student is blindfolded, spun around, and then the blindfold is removed and the student must grab as many moths as possible in five seconds. The easiest to see will be grabbed first!

Research: Wonders of Wonder Drugs

Some of the microorganisms that make people sick are becoming resistant to "wonder drugs" like penicillin. Challenge students to see how this is an example of natural selection leading to evolutionary change. This may be harder than thinking of insects or finches, because students do not seem to consider microorganisms to be alive in the same way that plants and animals are alive.

Mission 6

Natural Selection
Mission Briefing

Name:

Date:

Lori Morino, Biologist on the SETI Academy Team.

Some SETI scientists expect that if life began on another planet, the process of evolution might result in complex plants and animals (of course, the planet would need liquid water, an atmosphere, and a variety of raw materials to support life as we know it). The SETI project is trying to do an experiment to see if evolution on some other planets has already produced intelligent beings that are capable of sending out radio signals that we might detect.

To think about evolution on other planets, you need to explore how evolution occurred on Earth. Many different kinds of living things, called *organisms*, evolved on Earth. You have seen microorganisms, and you know about plants and animals. Each special kind of organism, like dogs, oak trees, or people, is called a *species*. Every species of plant or animal (or microorganism) has individuals that look and act differently from each other. These differences are often passed from parents to their children. For instance, you may have the same color eyes and hair as one or both of your parents. Your mission is to see if any of these differences might help a particular plant or animal have a better chance to live.

What Do You Think?

1. What kind of differences do you see among people? Among dogs?

2. Do you think these differences could be important to evolution? How?

From *How Might Life Evolve on Other Worlds?* © 1995. Teacher Ideas Press. (800) 237-6124.

Mission 6

Natural Selection

Directions for the Hungry Finch Game

1. Count out 50 of each kind of seed. Scatter them on the carpet.

2. One player uses chopsticks, another player uses salad tongs, and a third player uses a pair of pliers. Players are "finches" and these tools are their "beaks." A fourth player acts as the timekeeper. Any other players will start the game as seed counters.

3. The timekeeper should signal the start of a feeding frenzy. The three finches each try to pick up as many "seeds" as possible with their "beak" and place the seeds into their cup. Putting the cup on the carpet and shoving seeds into it is not allowed!

4. After 20 seconds, the timekeeper says, "Stop all feeding!"

5. Each finch should count the number of each kind of seed captured. Record the results into the "Data Table" logbook sheet.

    Any finch with *fewer* than 20 seeds has starved to death! That player is out of the game, and that player's particular beak difference has also died out.

    Any finch with 20 to 40 seeds continues playing the game.

    Any finch that has *more* than 40 seeds continues playing the game, *and* has a child! Add one more player with the same type of beak.

6. Empty all the cups. Put the "eaten" seeds into a pile on a desk.

7. The timekeeper should shake the remaining uneaten seeds off of the carpet and into a large plastic bag. The seed counters count the number of each kind of uneaten seed. Record these numbers into the "Data Table" logbook sheet.

8. The surviving seeds are allowed to reproduce by doubling their number. For example, if 15 green peas are left, add 15 more green peas. Get extra seeds from the teacher if you need them.

9. Scatter the surviving seeds and their "children" onto the carpet.

10. Repeat steps 3 to 9 for four more feeding frenzies. Be sure to record the information for each frenzy into the "Data Table" logbook sheet!

From *How Might Life Evolve on Other Worlds?* © 1995. Teacher Ideas Press. (800) 237-6124.

Mission 6
Natural Selection

Name:

Date:

As you play the Hungry Finch Game, record your data into this table.

Table 6.1—Data Table.

Feeding Frenzy	Predator Success—Number of Seed Picked Up	Prey Population—Number of Seeds Left on Floor
First	Chopsticks _____ Salad Tongs _____ Pair of Pliers _____	Brown Beans _____ White Beans _____ Green Peas _____
Second	Chopsticks _____ Salad Tongs _____ Pair of Pliers _____	Brown Beans _____ White Beans _____ Green Peas _____
Third	Chopsticks _____ Salad Tongs _____ Pair of Pliers _____	Brown Beans _____ White Beans _____ Green Peas _____
Fourth	Chopsticks _____ Salad Tongs _____ Pair of Pliers _____	Brown Beans _____ White Beans _____ Green Peas _____
Fifth	Chopsticks _____ Salad Tongs _____ Pair of Pliers _____	Brown Beans _____ White Beans _____ Green Peas _____

Mission 6

Natural Selection

Name:

Date:

As you play the Hungry Finch Game, record your data into this table.

Table 6.2—Data Table.

Feeding Frenzy	Predator Success—Number of Seed Picked Up		Prey Population—Number of Seeds Left on Floor	
First	Grabber	Number of Seeds	Kind of Seed	Number of Seeds
	_____	_____	_____	_____
	_____	_____	_____	_____
	_____	_____	_____	_____
Second	Grabber	Number of Seeds	Kind of Seed	Number of Seeds
	_____	_____	_____	_____
	_____	_____	_____	_____
	_____	_____	_____	_____
Third	Grabber	Number of Seeds	Kind of Seed	Number of Seeds
	_____	_____	_____	_____
	_____	_____	_____	_____
	_____	_____	_____	_____
Fourth	Grabber	Number of Seeds	Kind of Seed	Number of Seeds
	_____	_____	_____	_____
	_____	_____	_____	_____
	_____	_____	_____	_____
Fifth	Grabber	Number of Seeds	Kind of Seed	Number of Seeds
	_____	_____	_____	_____
	_____	_____	_____	_____
	_____	_____	_____	_____

Mission 6

Natural Selection

What Do You Think, Now?

Name:

Date:

After you have completed this mission, please answer the following questions:

1. Which beak types become *more* common in the finches? Why?

2. Which beak types become less common or were lost in the finches? Why?

3. Would a different color of carpet (different environment) change the results? Check with groups that used a different color of carpet.

4. How did the "seeds" change? Why?

Mission 7

A Timeline for the Evolution of Life
A Photo Album of Life on Earth

Overview

In mission 8 of the first book of the *SETI Academy Planet Project* (*The Evolution of a Planetary System*), students created a timeline of the geologic evolution of Earth. In mission 7.1, students see an image show depicting the major events in the *biologic* evolution of Earth, which reinforces the idea that, for most of Earth's history, the only life was microscopic In mission 7.2, students add pictures of the major events in the biologic evolution to the geologic timeline from *The Evolution of a Planetary System*, or they will create a new timeline specifically for biologic evolution. Larger, more complex, multicellular life-forms became numerous in an "explosion" about 730 million years ago. In geologic time, humans have existed for only a tiny fraction of Earth's lifetime.

Concepts

- In the fossil record, the oldest fossils are of simple microorganisms.

- During most of its history, life on Earth was microscopic.

- Complex life was numerous by 730 million years ago.

- The basic evolutionary sequence for vertebrates is: invertebrate to fish to amphibian to reptile to mammal/bird.

- People evolved very recently relative to Earth's geologic life span.

- The time period over which various species of organisms evolved on Earth is immense compared to a human lifetime.

Notes

In mission 6, students learned that natural selection can cause small changes in a species in a relatively short amount of time and that, given enough time, these changes can result in entirely new species.

- Factors that have shaped the evolution of life on Earth include environments and climates, catastrophic events that changed these environments, and random events such as the impact of a large asteroid (like the one widely believed to have landed in the Gulf of Mexico, causing the extinction of the dinosaurs), the collision of two drifting continents, or ice ages.

Skills

- Arranging evidence.

- Deductive reasoning.

- Measuring in metric units.

- Visualizing vast expanses of time.

Mission 7.1

Materials

For the Class

- *History of Earth* video

- VCR and monitor

- Timeline from mission 8 of the first book of the *SETI Academy Planet Project* (*The Evolution of a Planetary System*) or 5 meters of adding machine tape and meter sticks

- (optional) Transparencies of the black-line masters

- (optional) Overhead projector

- (optional) "Life Story of the Earth" video script

- (optional) Live fish, amphibians, reptiles, mammals, or birds (or pictures of these animals). One of each type (or one picture of each) will be best.

For Each Student

- SETI Academy Cadet Logbook

- Pencil

Getting Ready

One or More Days Before Class

1. (optional) Locate live animals to bring into the classroom. Perhaps ask students to bring in their small pets.

Just Before the Lesson

1. Set up the VCR and monitor or have the transparencies of the black-line masters ready. Start the *History of Earth* video at "Mission 7." Have the "Life Story of the Earth" video script handy.

Classroom Action

1. **Mission Briefing**. Have the class refer to the "Mission Briefing" for mission 7 in their student logbooks while one student reads it aloud.

2. **What Do You Think?** Have students answer the pre-activity questions on the "Mission Briefing." Invite them to share their answers in a class discussion.

3. **Lecture**. The image show that students will see assumes that they know what the terms *invertebrate, fish, amphibian, reptile, bird,* and *mammal* are. Go over definitions of these terms, especially *amphibian* versus *reptile*.

    (optional) Have students sort pictures of animals into the categories *invertebrate, fish, amphibian, reptile, bird,* and *mammal*.
    (optional) If live animals are available, discuss each one's classification and why each one is classified as it is. Ask students when they think that the first *kind* of fish appeared on Earth, according to the video images. Repeat this for any other categories of available live animals.

4. **Video**. Introduce the *History of Earth* video of still images. The "Life Story of the Earth" video script follows this lesson plan. Tell students that they are about to see an image show that recaps the major stages in the evolution of life on Earth. If the students saw the images in the first book of the *SETI Academy Planet Project*, tell them that the first time they saw the show, they watched for how the planet changed over time. This time, they are to think about the steps in the evolution of life on Earth.

Teacher's Note: Most people use the term animal *when they mean* mammal. *To be scientifically accurate, all* invertebrates *and* vertebrates *are* animals, but only fur-bearing, milk-producers like cows and people are mammals.*

Mission 7.2

Materials

For the Class

- 1 copy of the video images (from the black-line masters in the back of the book)

- 1 copy of the "Life Story of the Earth" video script

For Each Group of Students

- Colored markers or crayons

- Scissors

- Tape

- String

For Each Student

- SETI Academy Cadet Logbook

- Pencil

Getting Ready

One or More Days Before Class

1. Hang up the timeline from the *Evolution of a Planetary System*, or make a new timeline by following these directions:

   a. Hang the 5 meters of adding machine tape horizontally on a wall at the front of the classroom.

   b. At the far right edge of the tape, make a small mark and label it "TODAY."

   c. Put a mark every 10 cm along the adding machine tape, starting from the "TODAY" label.

   d. Make a large mark every 50 cm, and label the large marks "0.5 billion years ago," "1 billion years ago," "1.5 billion years ago," and so forth, all the way to "5 billion years ago" on the far left. (If students haven't been introduced to decimals, label the large marks "500 million years ago," "1,000 million years ago," "1,500 million years ago," and so forth, all the way to "5,000 million years ago" on the far left.)

2. Copy one set of the video images.

3. Make a copy of the "Life Story of the Earth" video script.

Classroom Action

1. **Activity**. Students will create a timeline. Divide the class into teams of two students each. Give one of the video images drawings to each team. If some drawings are left over, save them for early finishers.

   Tell students that a timeline is a good way to visualize the passage of long periods of time. Tell them that the class will complete (or create) the timeline of Earth's geologic history by adding significant events in the evolution of life on Earth. Point out that the video images drawings you have handed out are copies of the images from the video. Each drawing is like one photograph in the "Family Album of Earth." Each group is to color their drawing so it will show up from a distance, and tape it on the timeline at its proper position. Students should tape a string from the drawing to the point on the timeline when that change occurred.

   Call students' attention to the copy of the "Life Story of the Earth" video script, in which each drawing is briefly described. Students can use the script to find the correct ordering of images on the timeline. Invite questions and ask students to begin.

   Help students as necessary. Students who finish early can be given another drawing if there are any left. As students start to arrange drawings on the timeline, there will be a great many drawings near the period closest to "TODAY." Help students arrange the drawings so they can all be seen clearly and so they are all in chronological order.

Closure

1. **Review**. When the timeline is complete, have students gather around it and summarize what it shows. Point out the first fish, the first amphibian, and so forth.

2. **What Do You Think, Now?** Have students answer the post-activity questions on the logbook sheet "What Do You Think, Now?" Invite students to share their responses. Ask students

how their opinions have been changed by this mission.

3. **Preview**. Ask students how a fish could evolve into an amphibian. Tell students that, in the next mission, they will consider *descent with modification*—part of the evolutionary process—and find the answer to this question.

Going Further

Drawing Assignment: Future People

For a final picture on the timeline, have students speculate and draw their own concepts of the physical changes that might occur in humans in the next 5 million to 50 million years. What might happen if humans were able to colonize another planet with heavier gravity or a colder climate? Would they evolve differently than the people back on Earth?

Art Project: Illustrated Timeline

Have students use foam inner soles (from a shoe store or drugstore) or potatoes to create "rubber stamps" showing certain organisms, such as cyanobacteria. Have them stamp the timeline in each of the time periods that the organisms survived.

Public Speaking: Earth's Story

Have students write their own narration for the video image show and present it to another class.

Script for Video Images

"The Evolution of Life on Earth"

Figure 7.1—Earth, the third planet from the Sun, was special. Its orbit around the Sun was just right to allow Earth to have liquid water and to be comfortably warm.

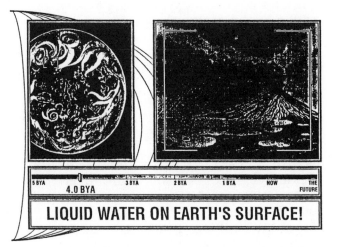

LIQUID WATER ON EARTH'S SURFACE!

Figure 7.2—Under these ideal conditions, very simple life appeared in the seas of Earth about four billion years ago. How and where did life begin? This is still one of the greatest mysteries in science. This image shows two theories. One is that life first formed near the ocean's surface or along the shore, using energy from lightning or from the Sun. The other is that life formed at the bottom of the ocean, with energy from volcanic vents. No one knows for sure if either of these theories is correct.

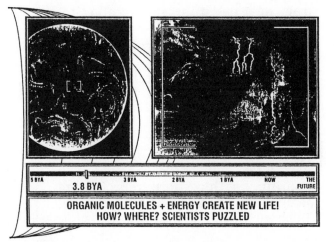

ORGANIC MOLECULES + ENERGY CREATE NEW LIFE!
HOW? WHERE? SCIENTISTS PUZZLED

Figure 7.3—What is fairly certain is that tiny microorganisms called *bacteria* somehow came into existence. They dominated the seas for over two billion years, and their descendants still live all over the world today. The first bacteria lived in an atmosphere that was high in carbon dioxide and very low in oxygen. Like us, they had to eat food. They ate the organic molecules in the sea. But as this food ran out, a new type of organism that could make its own food emerged. It used carbon dioxide and released oxygen.

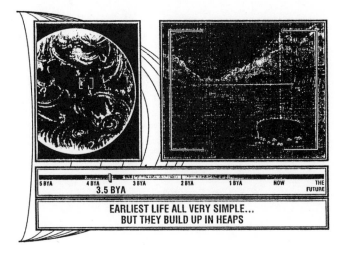

EARLIEST LIFE ALL VERY SIMPLE...
BUT THEY BUILD UP IN HEAPS

Figure 7.4—This new type of organism was a *cyanobacteria*. As many cyanobacteria made their own food, they used carbon dioxide and released oxygen. Life itself was changing the Earth's atmosphere! For millions of years, oxygen was slowly added to the atmosphere.

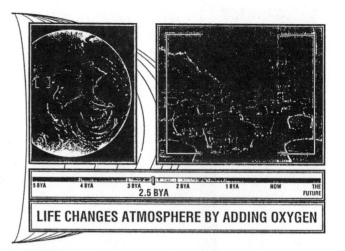

Figure 7.5—During this time, simple one-celled organisms reproduced by splitting apart to form two identical cells that were exactly alike. But about 1.7 billion years ago, some organisms began to reproduce by joining together and creating an entirely new cell that was similar to, but not exactly like, the parent cells. This was the beginning of *sexual reproduction*, a real boost to variety among new organisms. This meant many more differences for natural selection to work on.

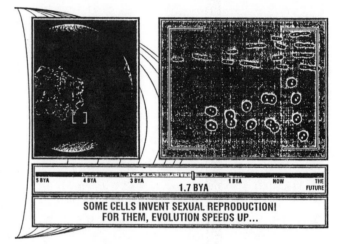

Figure 7.6—Seven hundred million years later, some cells that failed to separate as they were reproducing formed many-celled living organisms. These included the first seaweeds.

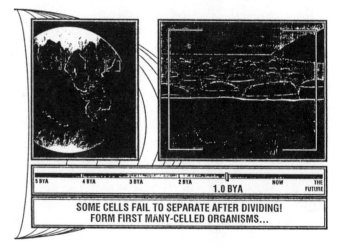

Figure 7.7—Some animals eventually developed the first specialized cells that worked together to form simple nerves and muscles. These animals had some control of their movement. Some of these ancient animals were like the planaria that still live today.

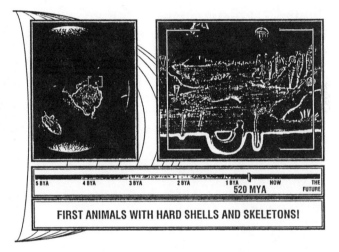

ANIMALS EVOLVE SPECIALIZED CELLS! NERVES, MUSCLES...

Figure 7.8—Another 200 million years later, soft-bodied animals and plants in the seas were joined by some fascinating-looking creatures with hard outside shells and skeletons. These were animals like the trilobite and the ancestors of snails. Both soft-bodied and hard-bodied animals with no backbone are called *invertebrates*.

FIRST ANIMALS WITH HARD SHELLS AND SKELETONS!

Figure 7.9—As more oxygen was produced, Earth's atmosphere developed an *ozone layer* that screened out some of the Sun's harmful ultraviolet rays. Organisms could finally survive on land. Bacteria and plants were the first to leave the sea for the land. The seas continued to be full of invertebrates, including new kinds. The first animals with backbones, called *vertebrates*, were primitive kinds of fish. Many, many kinds of fish swam in the ancient seas.

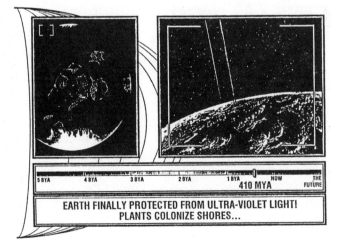

EARTH FINALLY PROTECTED FROM ULTRA-VIOLET LIGHT! PLANTS COLONIZE SHORES...

Figure 7.10—Some of the fish began to live in fresh water. One type of freshwater fish, called the *lobe fin*, eventually developed the ability to walk from pond to pond, probably because some ponds were drying up. One of the kinds of lobe fin fish evolved into the four-footed *Ichthyostega*, a primitive amphibian who was surrounded by lots of insects and plants in its new home on land.

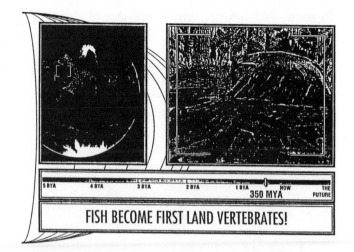

FISH BECOME FIRST LAND VERTEBRATES!

Figure 7.11—By about 300 million years ago, amphibians, huge insects, giant centipede-like animals, and the first scaly reptiles inhabited a land with giant ferns and other plants. As these plants died, their decaying remains formed the world's coal deposits, resources we still use today. Do you see why coal is called "fossil fuel"? Life was lush and abundant.

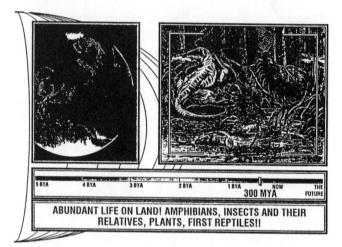

ABUNDANT LIFE ON LAND! AMPHIBIANS, INSECTS AND THEIR RELATIVES, PLANTS, FIRST REPTILES!!

Figure 7.12—About 225 million years ago, Earth's greatest mass extinction occurred. Was this major loss the result of an asteroid impact? A nearby exploding star? A giant volcano? Scientists do not know. Whatever the reason, 96 percent of the sea life, 75 percent of the amphibians, and 80 percent of all land species vanished. Such catastrophes have been a very important factor in the evolution of life on Earth. The extinction of dominant species meant that the surviving species could evolve further and expand their populations.

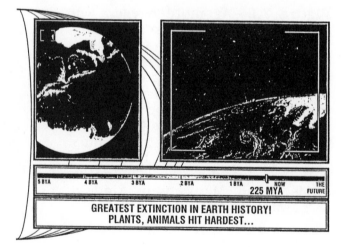

GREATEST EXTINCTION IN EARTH HISTORY! PLANTS, ANIMALS HIT HARDEST...

Figure 7.13—Some small reptiles survived the catastrophe and continued adapting until they had evolved into new kinds of reptiles. Some became dinosaurs and others became mammals. Some reptiles even developed the ability to fly.

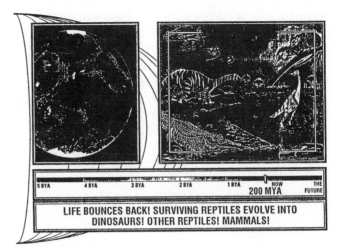

LIFE BOUNCES BACK! SURVIVING REPTILES EVOLVE INTO DINOSAURS! OTHER REPTILES! MAMMALS!

Figure 7.14—Then, about 65 million years ago, there was another large extinction. The cause? We have evidence that this time it was probably an asteroid that struck the Earth in the Gulf of Mexico. This was the time the dinosaurs became extinct. However, some organisms, including some small reptiles, small mammals, birds, and fish, did survive. If it were not for this catastrophe, Earth might still be dominated by dinosaurs rather than by people! Or, *maybe* we would be here together!

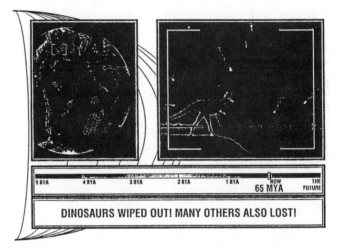

DINOSAURS WIPED OUT! MANY OTHERS ALSO LOST!

Figure 7.15—Some small reptile-like mammals adapted so well that they increased dramatically in variety and number. They evolved into true mammals. Some became large. Most of the mammals that exist today can trace their ancestors to this period of time.

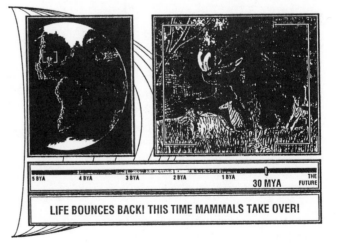

LIFE BOUNCES BACK! THIS TIME MAMMALS TAKE OVER!

Figure 7.16—About two to five million years ago, one group of mammals began to use natural tools, and later to make tools of their own, to use fire to keep warm, and to cook their food. These major steps led early human-like animals, called *hominids*, to control their environment, settle in groups, and form communities. Eventually, just a few hundred thousand years ago, our species of hominids, called *Homo sapiens*, or humans, migrated to nearly all parts of Earth.

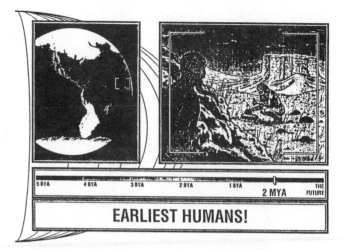

EARLIEST HUMANS!

Figure 7.17—From these early human communities came a variety of cultures that have produced the young people of today. As a member of this group, you are part of a new stage in Earth's development—one that has radio, television, space satellites, and computers as tools that can be used to help people understand each other better and answer questions that humans have been asking for centuries: Are we alone? Or are there other intelligent beings out there somewhere on other planets in other solar systems who are as anxious to find out about us as we are to find out about them?

NOW! CHILDREN OF THE EARTH HERE!

Figure 7.18—We don't know what changes evolution will bring about in the physical characteristics of people in the distant future, but natural selection will continue. How do you think people will be different five million years from now?

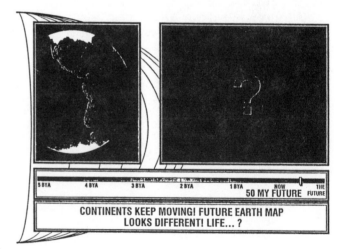

CONTINENTS KEEP MOVING! FUTURE EARTH MAP LOOKS DIFFERENT! LIFE... ?

Mission 7

A Timeline for the Evolution of Life

Mission Briefing

Name:

Date:

Victoria Johnson, Biologist on the SETI Academy Team.

To see the vast amount of time that has been available for natural selection to act and to create new species of animals, you will create a timeline showing the major events in the evolution of life on Earth. You will use a scale of 1 meter = 1 billion years (1,000 million years) to complete the timeline. To begin, you need to consider how animals are classified, and you need to see an image briefing about animal evolution.

What Do You Think?

1. If you displayed all of Earth's history, from 4.6 billion years ago to now, on a 4.6 meter timeline, how many meters or centimeters would you have to measure from the present to the point when humans first appeared?

2. How far back from the present, in meters or centimeters, would simple microscopic life begin?

Figure 7.19—Earth's History Timeline.

5 bya	4 bya	3 bya	2bya	1 bya	Now
5 meters	4 meters	3 meters	2 meters	1 meter	

Mission 7

A Timeline for the Evolution of Life

Table 7.1—A Timeline for the Evolution of Life.

Image	Time	Forms of Life
7.1	4.0 bya	No life; shallow seas
7.2	3.8 bya	Origin of simple cells
7.3	3.5 bya	Origin of cyanobacteria
7.4	2.5 bya	Oxygen accumulates in atmosphere
7.5	1.7 bya	Protists and green algae
7.6	1.0 bya	Simple multicellular life (sponges, seaweeds)
7.7	700 mya	More invertebrates (flatworms, jellyfish)
7.8	520 mya	Early animals with hard parts in oceans
7.9	410 mya	Plants invade land
7.10	350 mya	Vertebrates invade land
7.11	300 mya	Coal forming forests, amphibians, BIG insects
7.12	225 mya	Mass extinction (trilobites)
7.13	200 mya	Pangaea, first mammals, first reptiles
7.14	65 mya	Mass extinction, including dinosaurs
7.15	30 mya	Small mammals, humanoids
7.16	2 mya	Early man
7.17	0 mya	Us
7.18	50 mya forward	What will Earth be like in 50 million years?

Mission 7

A Timeline for the Evolution of Life
What Do You Think, Now?

Name:

Date:

After you have completed this mission, please answer the following questions:

1. During how much of Earth's history was microscopic life present?

2. How recently did mammals evolve?

3. How recently did hominids and people evolve?

4. If we discovered another planet suitable for life that had developed in a manner similar to the way Earth developed, is it more likely that we would find simple life-forms or complex life-forms there? Why?

5. Draw and describe some simple and complex life-forms that might have evolved on another planet.

Mission 8

Tracing Family Trees
Life Is Such a Puzzle!

Overview

In mission 8.1, students take part in a discussion about evolution and the work of paleontologists. They construct a recording sheet showing geologic periods. In mission 8.2, students see *descent with modification* as they attempt to trace the family trees of trilobites and the major kinds of vertebrates. They get a glimpse into one of the exciting puzzles of science—interpreting the fossil record to trace possible lines of descent—as they arrange pictures to show how each kind of animal may have evolved from a previous kind of animal. In this mission, students learn that some species became extinct millions of years ago. In fact, there are many more extinct species than there are living species.

Notes

In mission 6, students studied natural selection, the mechanism of evolution. In mission 7, students made a timeline showing the evolution of life on Earth. The timeline reflected some of the general trends in evolution, such as the fact that single-celled creatures precede multicelled creatures.

Concepts

• Complex life-forms evolved from simpler life-forms.

• Certain biologic concepts apply to all animal life-forms.

• There are many similarities among Earth animals, especially among the vertebrates. These similarities are often the result of common descent from a shared ancestor.

• Today's vertebrates all have a common ancestor.

• Structures may evolve from some other, earlier structure by natural selection. Darwin called this *descent with modification*.

Skills

• Arranging evidence.

• Visualizing the evolution of life.

119

- Forming hypotheses.

- Deductive reasoning.

Mission 8.1

Materials

For the Class

- Transparencies of a sample recording sheet (see fig. 8.1, page 124) and Fossil pictures I, II, and III (figs. 8.3, 8.4, 8.5 on pages 129, 130, 131).

- Overhead projector

- (optional) Chicken, cat, or frog skeleton

For Each Pair of Students

- Scissors

- Transparent tape

- 2 large (18-by-24-inch or larger) sheets of construction paper

- (optional) Marking pens, colored

- Ruler or meter stick

- Envelope

For Each Student

- SETI Academy Cadet Logbook

- Pencil

Getting Ready

One or More Days Before Class

1. Make transparencies of the sample recording sheet and the sample animals. Cut out the animals so that they can be moved around on the transparency of the recording sheet.

Classroom Action

1. **Mission Briefing**. Have the class refer to the "Mission Briefing" for mission 8 in their student logbooks while one student reads it aloud.

2. **What Do You Think?** Have students answer the pre-activity questions on the "Mission Briefing." Invite them to share their answers in a class discussion.

3. **Lecture**. The following information is offered as a thumbnail sketch of a lecture for students on how organisms evolve. We often talk about one species evolving into another, but we seldom consider how "new" features, such as eyes or legs, arise. Darwin proposed the concept of *descent with modification*. Natural selection takes an existing structure and may improve the design or may modify it to fit a new, different function. The most dramatic examples involve evolution of "complex" structures from earlier "simple" structures, such as eyes from single, light-sensitive cells.

   Table 8.1 (see page 122) gives some examples of features of Earth organisms, and the structures from which they evolved. This table of examples is not exhaustive, but it illustrates that the structures of organisms are traceable to earlier, simpler structures in their ancestors. In these cases, the connection is very well documented by fossils, embryology, and comparative anatomy. Every structure evolves from some other preexisting structure. Much of evolution also includes the evolution of a "complex" structure from some preexisting feature that was just as complex as the new feature (a bat's wing from a mammal's front leg), or even the evolution of "simpler" structures from preexisting structures that appeared more complex (a sea lion's front flipper from a mammal's front leg). Natural selection works with whatever raw material is available.

4. **Discussion**. Tell students that the fossil record is very extensive, but it is not complete. Very few organisms become fossils, and, even of those that do, we probably have not found very many. When paleontologists want to interpret the fossil record to trace possible lines of descent, they must make hypotheses based on the clues that they do have.

   Tell students that they will receive pictures of animals that were reconstructed from fossils. (Fossil Pictures I, II, and III from the logbook sheets.) These pictures are not to scale! They have been reduced to fit into the space provided.

The pictures that students are given each represent years of painstaking research—first finding and excavating the fossils, fitting broken pieces together, and reconstructing what the animal probably looked like by comparing its bones with those of modern creatures.

Table 8.1—Descent with Modification.

Feature	Structure from Which It Evolved
Eyes	Simple, light-sensitive cell, then lots side by side
Limbs	In vertebrates, from fins on fish-like ancestors
	In insects and centipedes, from flaps on side walls of worm-like ancestors
Wings	In vertebrates, from forelimbs of reptiles or mammals
	In insects, from stiff flaps extended from body wall
Poison fangs	In snakes, grooved fangs became hollow fangs; salivary glands secrete toxin
	In centipedes, spiders, "fangs" are stiff, sharp front limbs connected to salivary glands
Armor	In turtles, expanded rib bones fused to hardened skin
	In armadillos, expanded hard lumps in skin form armor
Horns	Growths of bone from skull covered by hardened skin (antelopes, cattle), groups of hard, stiff hairs (rhinos), bony growths after skin falls off (deer); hardened exoskeleton (beetles); shell secreted by glands in skin (molluscs)
Stings	Hardened, sharp ovipositor in neuter (female) worker bees and wasps; hard exoskeleton spine (scorpions)
Hooves	Enlarged, hardened toenails
Tusks	Elongated teeth (elephant, saber-tooths, walruses)
Feathers, hair	Modified reptilian scales
Electric shock ability	In electric eel; exaggeration of ability of some fish to use electric fields for navigation; electricity is generated by activities of many cells (for example your heart, as seen by EKG); these fish have enlarged arrays of these cells
Lungs	In vertebrates, modified swim bladder of fish ancestor (which was a gas-filled space used for buoyancy in water)
	In land snails/slugs, the space that once held the gill of the aquatic ancestor, now lined with lots of blood vessels
Eardrums	Modified small bones that once worked the jaw joint in fish; in insects a modified plate of the exoskeleton
Jaws	In vertebrates, the bent, enlarged, bony joints that once supported a gill
	In insects, centipedes, spiders, crustaceans, mouthparts are modified front limbs that were once on the front end of a more centipede-like ancestor

(optional) Show a chicken, cat, or a frog skeleton. Have students see how the bones fit together like a jigsaw puzzle. Have them imagine what these animals looked like while they were alive, based upon the skeletons. They should get a general idea, but not details like color or hair length.

Tell students that we have a good idea how each fossilized creature looked when it was alive. This is one good clue that a fossil provides. How does this help us find probable lines of descent? We can imagine that structures in one form evolved into somewhat different structures in the newer animal. Simple fins come first, then jointed "lobe fins," then simple legs, then better legs. Legs can even evolve into wings! We can see a sequence of changes leading from one form to another.

There is a second good clue to ancestry that fossils provide. What could it be? The *age* of the fossil. If two fossils look similar, like two dinosaurs or two trilobites, one kind could be the ancestor of the other kind. If one kind of fossil is older than the other, the older one should be the ancestor; the younger kind could not be the ancestor of the older one!

5. **Demonstration**. Show students how to build a family tree. They will use the two clues—visible structures and the age of the fossils—to trace two lines of descent. First they should see how separate these two lines are.

Ask students to think about all the kinds of animals they know about that are alive today. Many of these animals have backbones. Animals with backbones are called *vertebrates*. People are vertebrates. So are dinosaurs, buffalo, and snakes—but not worms. There are animals without backbones: worms, sea stars, clams, and lots of others. Animals without backbones are called *invertebrates*. This is a significant difference. It is likely that vertebrates and invertebrates represent two major lines of evolutionary descent. Vertebrates will evolve into more vertebrates, and invertebrates will evolve into more invertebrates.

Turn on the overhead projector. Place the recording sheet transparency on the projector. See figure 8.1. Explain to students that this image is like the recording sheet they will make. Place the recording sheet so that the "50 mya-present" end is close to the floor (the bottom of the projector screen). Tell students that the bottom represents the oldest known layer of soil. Refer back to asking students to imagine that they were paleontologists. If necessary, review sedimentary rock and fossil formation.

Explain to students that they will turn the recording sheet sideways when they use it to make family trees as shown in figure 8.6 (page 133). Point out that this does not change the geologic age of any fossil found in any of these time periods.

Figure 8.1—Recording Sheet.

	50 mya - present
	100 mya - 50 mya
	150 mya - 100 mya
	200 mya - 150 mya
	250 mya - 200 mya
	300 mya - 250 mya
	350 mya - 300 mya
	400 mya - 350 mya
	450 mya - 400 mya
	500 mya - 450 mya
	550 mya - 500 mya
	600 mya - 550 mya

6. **Activity**. Divide students into teams of two. Ask a volunteer to read the "Making a Recording Sheet" logbook sheet while the class reads along. Have each student pair create a recording sheet. Make sure that they label all the geologic time period headings correctly onto the columns of their recording sheets. Circulate to make sure every group has their recording sheet showing oldest to most recent going from left to right across the sheet, and that student names are on the recording sheet. When the recording sheets are complete, allow student pairs a few minutes to cut the fossil reconstruction pictures apart.

(optional) If there is time, have students color-code the pictures: perhaps red for trilobites, green for amphibians, and so forth. Have students place their animal pictures into envelopes for safekeeping until mission 8.2. Collect their envelopes and recording sheets.

Mission 8.2

Materials

For the Class

- Overhead projector

- Transparency of a sample recording sheet

- Transparencies of sample animal cutouts from the "Fossil Pictures" logbook sheets

For Each Pair of Students

- Transparent tape

- (optional) Marking pens, colored

For Each Student

- SETI Academy Cadet Logbook

- Pencil

Getting Ready

Just Before the Lesson

1. Distribute or have student pairs collect their family tree materials from mission 8.1.

Classroom Action

1. **Demonstration**. Each team will be asked to arrange the animal pictures to show how the kinds of vertebrates in the pictures may have evolved from earlier kinds of vertebrates and how the kinds of invertebrates in the pictures may have evolved from earlier kinds of invertebrates.

    Place the recording sheet on the overhead projector. The pictures need to be placed into their appropriate time-period columns, and then organized in these columns so they make a family tree that connects animals that may have evolved into other animals. Use the cutout animal pictures to demonstrate the next steps on the overhead projector, as shown in figure 8.2. Do not show students the whole picture!

    a. Sort the organism pictures into the various time periods, such as the 500-450 mya time period.
    b. Set the organism pictures into their appropriate columns.
    c. Compare the fossil pictures in each pair of two adjacent columns. Ask students which ones could be related. The trilobites are obviously related to (are ancestors or descendants of) other trilobites. The fish-shaped animals are less obvious, but they do form a sequence of vertebrate descendants.
    d. "Tape" a couple of sample trilobites side by side at the bottom of the page into their appropriate columns, and connect them with a line. Do the same for the "fish" but put them at the top of the page.

    Recap any of the relevant discussion about relationships among fossils from different time spans: Look for number of appendages—legs, wings, tails, and fins— to determine relatedness. Remind students about *descent with modification*. They may see that vertebrates did not evolve any "new" appendages after the 400-350 mya time period, but they did change the usage of the ones they already had, like fins to legs or legs to wings. They may see that the basic body shape often stays consistent, although body covering may change (scales to feathers or fur).

Teacher's Note: We are not *saying that the exact species shown in one picture evolved into the exact species shown in the next picture. We are saying that some kind of fish evolved into some kind of amphibian. Paleontologists would be able to be more specific, as they have many more fossils to examine! Make this point clear to students.*

Figure 8.2—How to Connect Family Lines.

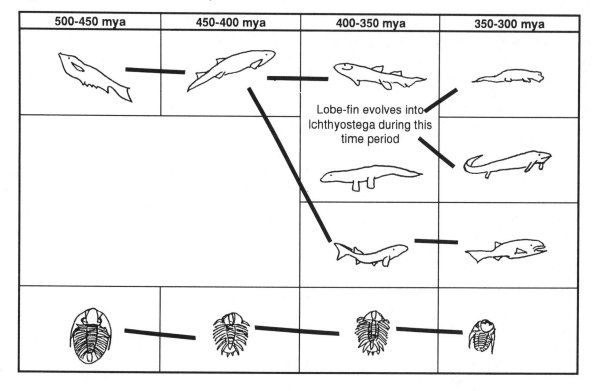

Tell students that finding relationships among ancient organisms, based only on fossils, is very difficult. They will experience some of the frustrations that paleontologists experience in their work. The expectation for this assignment is not that they turn in a perfectly plausible vertebrate family tree but that they complete the assignment based on their best guesses about relationships among related organisms. They will be expected to explain how they decided which life-forms might have evolved from earlier life-forms.

2. **Activity**. Return the envelopes with the cut-out pictures and the recording sheets to students. Allow time for students to build their family trees, circulating and helping as needed.

   (optional) If students are still confused about drawing lines to trace the evolving animal family trees of fish, amphibians, reptiles, birds, and mammals, they can color-code the animal pictures (if they have not already done so). For example, color all amphibians with shades of green.

Closure

1. **Discussion**. Have teams of students compare their family trees with others. If the trees are different, ask students why. Scientists don't usually agree until they have discussed all the possibilities and tested alternative hypotheses. Conduct a class discussion. Ask the following questions, to be answered based upon class data.

   - Which kind of modern animal appeared first? Last? In the middle?
   - When did branches develop in the family tree? What did they produce?
   - When was the number of different kinds of animals alive the smallest? The largest?
   - What does the mammal family tree trace back to? All vertebrates, fish, amphibians, reptiles, birds, mammals?
   - What are some animals that were alive in the past that are not alive now?
   - What are some animals that are alive now that were not alive in the past?

2. **What Do You Think, Now?** Have students answer the post-activity questions on the logbook sheet "What Do You Think, Now?" Invite students to share their responses. Ask students how their opinions have been changed by this mission.

Going Further

Research: Other Family Trees

Is the lowly chicken a descendant of the mighty *Tyrannosaurus rex*? Yes—in a manner of speaking. Certain dinosaurs evolved into birds, one modern example of which is the chicken. Have students create a family tree of the chicken, cat, dog, horse, or any other creatures.

Mission 8

Tracing Family Trees
Fossil Pictures

Figure 8.3—Fossil Pictures I.

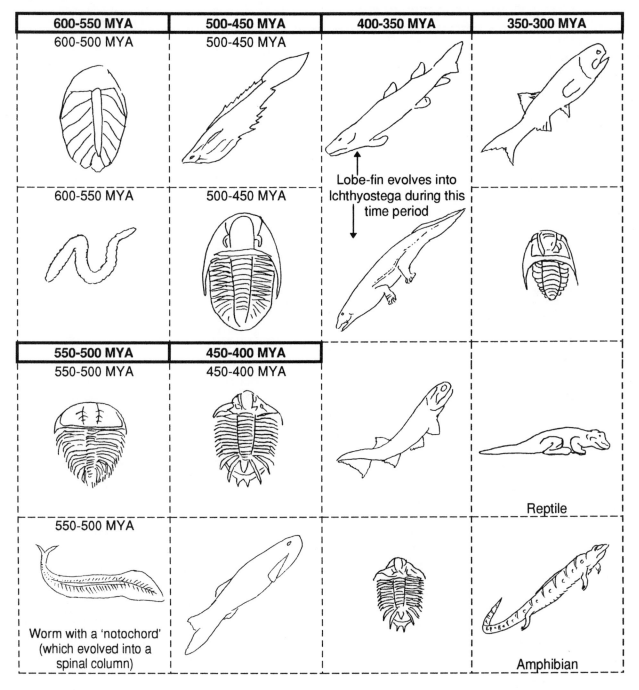

From *How Might Life Evolve on Other Worlds?* © 1995. Teacher Ideas Press. (800) 237-6124.

Mission 8

Tracing Family Trees

Fossil Pictures

Figure 8.4—Fossil Pictures II.

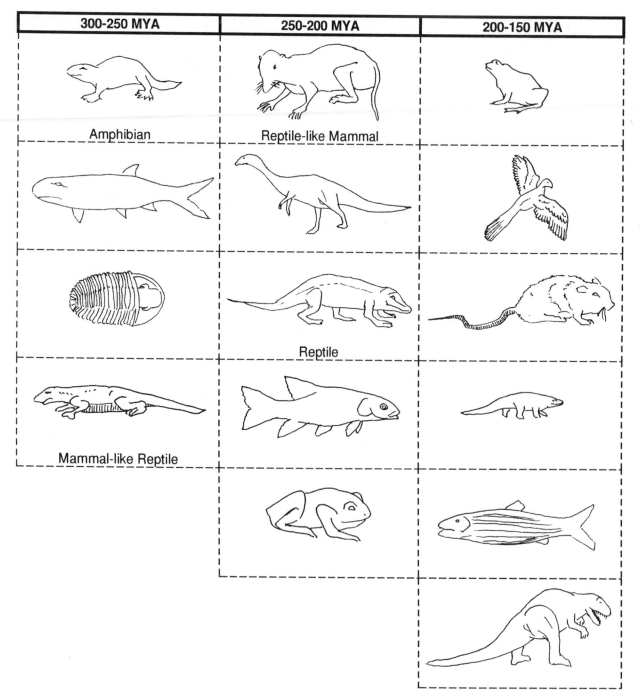

300-250 MYA	250-200 MYA	200-150 MYA
Amphibian	Reptile-like Mammal	
	Reptile	
Mammal-like Reptile		

Mission 8

Tracing Family Trees

Fossil Pictures

Figure 8.5—Fossil Pictures III.

150-100 MYA	100-50 MYA	50 MYA - Present

From *How Might Life Evolve on Other Worlds?* © 1995. Teacher Ideas Press. (800) 237-6124.

Mission 8

Tracing Family Trees
Mission Briefing

Name:

Date:

Dr. Rufus Catchings, Paleontologist on the SETI Academy Team.

Most of the kinds of animals that have ever walked or flown or swam on Earth evolved in the last 730 million years. How do we know about them? The best evidence is from the careful digging, dating, and sorting of fossils.

In this mission, we would like you to examine several drawings of trilobites and of vertebrates (animals with backbones) that were reconstructed from fossils. Suppose that these pictures showed the only fossils that you have been able to collect. Given information about the period of time when these animals lived, we would like you to hypothesize how they, or vertebrates very much like them in appearance, may have evolved from earlier kinds of animals. You will create a family tree for them that shows their evolution to a few of the new species alive today.

What Do You Think?

1. Evidence suggests that simple worm-like organisms were the first living animals to appear on Earth. How do you think these organisms might have evolved into all of the multicellular animals that exist on Earth today?

2. Even further back in time, simpler creatures like bacteria, blue-green bacteria, and protists existed. Were some of these the ancestors of all of the multicellular plants and animals that exist on Earth today?

Mission 8

Tracing Family Trees

Making a Recording Sheet

1. Take two large pieces of construction paper and make six equal columns on each page. (You can do this by folding each sheet in half and then in thirds, or by using a ruler and pencil to mark off each sheet into six equal columns.)

2. Tape the two papers together to create one large recording sheet, as shown in fig. 8.6.

3. Label each column with a geologic time period. Put the oldest time period on the left, and the time period that is closest to the present on the right.

4. Cut out the drawings of fossil animals on the three "Fossil Pages" logbook sheets.

5. Color these cutout fossil animals if you wish.

6. Put the drawings into an envelope, put your groups' names on the envelope, and give it to your teacher. In the next activity, you will rearrange these pictures to show your family trees.

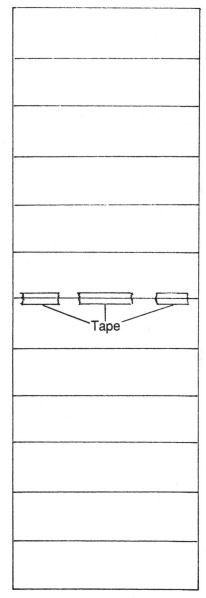

Figure 8.6—Making a Recording Sheet.

Tape

Mission 8

Tracing Family Trees

Building Family Trees

1. Sort the fossil animal drawings into one pile for each time period.

2. Put all the fossil pictures into their correct time-period column.

3. Now you are ready to rearrange the pictures to show your family trees. When you arrange the pictures, keep in mind that any one kind of animal may either a) stay the same from one time period to the next, b) die out (become extinct), or c) evolve into different organisms that look somewhat like the original.

4. Move the pictures around until their ordering across time periods makes sense.

5. When you are satisfied with your family tree, tape the pictures to the recording sheet, and draw lines connecting animals that evolved into other animals. For example:

Figure 8.8—Connecting Pictures to Show Evolutionary Lines.

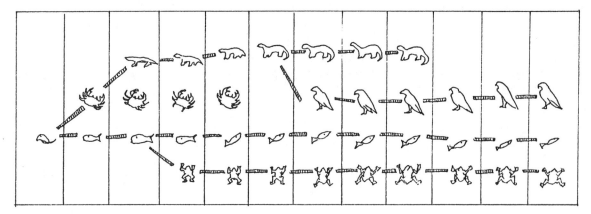

Mission 8

Tracing Family Trees

What Do You Think, Now?

Name:

Date:

After you have completed this mission, please answer the following questions:

1. How do you decide which life-forms are most similar to earlier forms of life? What clues do you look for?

2. Select one of the vertebrates that is alive today. Use your family tree to explain how you think this vertebrate might have evolved from earlier vertebrate life-forms.

Mission 9

What Organism Do You See?
Classifying Organisms by Characteristics

Overview

In mission 9.1, students learn how a dichotomous classification key works as they divide 20 different objects into consecutively smaller groups. In mission 9.2, students play a game called, What Organism Do You See?, that leads them through using the dichotomous classification key to identify what organisms have "attacked" them (been taped on their backs). The game introduces students to a few of the similarities among Earth organisms, and helps them start thinking about how life-forms on other planets may have evolved differently. The more students learn about the variety of such characteristics of Earth organisms, the more creative they will be in imagining possible organisms on other planets.

As far as scientists know, the process of evolution will operate on other planets as it has on Earth. However, if there is life on other planets, it likely will not be life as we know it, because natural selection depends on environment, which will probably be different on other planets. And chance always plays its role.

Concepts

- There are five kingdoms of life on Earth: plants, animals, protists, bacteria, and fungi.

- There are many similarities and differences among Earth plants and animals.

- Biologists use these similarities and differences to classify Earth plants and animals.

- These similarities and differences are all adaptations to specific environments, which are caused by natural selection.

- Unfamiliar plants and animals can be identified by looking at their similarities and differences to known plants and animals.

Notes

In mission 8, students made family trees based upon the idea of descent with modification *and the fossil record.*

- Because environments on other planets are likely to be different from that on Earth, life on other planets is likely to evolve into different forms than it did on Earth.

Skills

- Using a dichotomous key.

- Inductive reasoning.

Mission 9.1

Materials

For the Class

- 10 to 20 types of common objects for sorting, such as a paper clip, a button, and an eraser

For Each Student

- SETI Academy Cadet Logbook

- Pencil

Getting Ready

Just Before the Lesson

1. Collect the 10 to 20 common objects for sorting onto a tray in plain view.

Classroom Action

1. **Mission Briefing**. Have the class refer to the "Mission Briefing" for mission 9 in their student logbooks while one student reads it aloud.

2. **What Do You Think?** Have students answer the pre-activity questions on the "Mission Briefing." Invite them to share their answers in a class discussion.

3. **Lecture**. Tell students that, as far as scientists know, the process of evolution will operate on other planets as it has on Earth. However, if there are organisms on other planets, they likely will not be the same as organisms on Earth, because natural selection depends on environment, which will certainly be different

on other planets. Natural selection also depends on chance events.

On Earth, successful characteristics evolved when chance variations turned out to be well adapted to the environment: In plants we see green leaves, root systems, and flowers; in animals we see exoskeletons, internal skeletons with backbones, and skin coverings of fur, feathers, and scales. Biologists use these structural similarities as a means of classifying animals and as a means of inferring their common evolutionary origins.

Have students start by imagining that they are biologists from another star system that have landed on Earth this year. Previous biologists have identified and described 16 different plant or animal species in this area. Tell students that, as they emerge from their spaceship, organisms jump on some of their backs! Afraid to remove them, they ask their partners questions to determine what the organisms are. Tell students that they will use the "Organism Classification Key" in their logbooks to help them ask intelligent questions. Have students try to identify as many organisms as they can.

4. **Activity**. Tell students that before they play the game called What Organism Do You See?, you would like them to help with a practice game that will make them aware of the need for careful observations to scientifically identify and classify organisms according to their likenesses and differences. Guide students through the following activity to develop a dichotomous key.

Have students sit in a circle on the floor or outside. Place the 10 to 20 common objects for sorting in the center of the circle. Give a brief explanation of the term *dichotomous*— double branching—to differentiate this key from other types of keys, like the "Organism Classification Key" in their logbooks.

Ask a student to identify and separate the items into two groups based on one observable characteristic, such as color or shape, as shown in figure 9.1. For example, the first group may be all round and curved objects, while the second group is all square and straight objects. Record students' description or reason for the groupings on the chalkboard.

Figure 9.1—Dichotomous Key I.

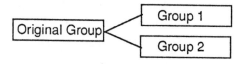

Ask another student to create two smaller groups from one of the branches created by the first student, as shown in figure 9.2. Record students' description or reason for the groupings on the chalkboard.

Figure 9.2—Dichotomous Key II.

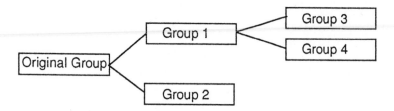

Continue to have students subdivide the branches until each of the items have been separated from all of the others. Discuss which descriptions were clear, which were too vague, which were too lengthy, and which were easily understood. Take a new item, such as a piece of chalk, and see if it will follow a path to its own separate, logical destination point. Follow the branches in reverse order, putting the items back together so that likenesses become evident.

(optional) Ask students if there is another way in which a dichotomous key could be created for these same objects; have students create this new key.

Mission 9.2

Materials

For the Class

- 4 copies of the organisms logbook sheets (5 pages)

For Each Student

- SETI Academy Cadet Logbook

- 2 strips of masking tape per student, 5 cm long

- Sheet of scratch paper

- Pencil

Getting Ready

Just Before the Lesson

1. Make 4 copies of the organisms logbook sheets.

2. Cut these apart, into strips, or plan to have students cut them apart.

Classroom Action

1. **Lecture**. Tell students that they are about to play a game called What Organism Do You See? Explain the rules as follows:

   a. Students form teams of two or three biologists.
   b. Each team receives one strip of paper for each student in the team. Each strip of paper describes an Earth organism. The papers are placed face down so they cannot be seen.
   c. The biologist who is "attacked" first has one strip of paper taped to their back so that it can only be seen by teammates.
   d. The attacked biologist uses the "Organism Classification Key" logbook sheet to ask questions about the organism until its identity is guessed. Teammates then remove the organism and show it to the attacked biologist.
   e. Teammates take turns playing the part of the biologist who is attacked. A team that has guessed the identities of all of their organisms should get another set from their teacher or trade with another team. The object of the game is for the team to guess as many organisms as possible during the class period.

2. **Discussion**. Allow time for questions.

3. **Activity**. Form teams of two or three players each and hand out the organism strips to each team face down. Allow the students to play for 15-20 minutes, or until all students have had a chance to identify at least two organisms.

Closure

1. **Discussion**. Ask teams how many organisms they identified successfully. No reward need be given other than the achievement of identifying a lot of organisms correctly.

2. **Lecture**. Tell students that the classification key is a simplified version of questions that real biologists ask whenever they encounter a new organism. By inspecting the organism to determine the answers to such questions, they can classify the organism. Biologists use many official divisions for classifying life on Earth.

3. **Review**. Ask students where individual characteristics come from. Perhaps review the theory of evolution, which explains how today's plant and animal species evolved.

4. **Activity**. Allow time for students to create a list of characteristics of plants or animals on Earth.

5. **What Do You Think, Now?** Have students answer the post-activity questions on the logbook sheet "What Do You Think, Now?" Invite students to share their responses. Ask students how their opinions have been changed by this mission.

Teacher's Note: There are two types of keys: artificial and taxonomic. Artificial keys are based upon any superficial characteristics. Taxonomic keys are based upon structures that are actually used in the classification of animals, such as "backbone" or "no backbone." These are often the structures that are derived from a common evolutionary ancestry. In this mission, students made artificial keys.

Going Further

Activity: Follow the Winding Road

Have students study flowcharts and the ways they can explain processes and come to solutions through flowcharts. Explain that the instrument used in this mission is a type of flowchart. Explain a simple card game like UNO® or Go Fish!® with a flowchart.

Activity: Make Personal Flowcharts

Have students make a flowchart for their daily activities. For example: Wake up. Eat breakfast. Go to school. Is it Tuesday? If yes, go to gym; if no, go to music. Lunch. Is it Monday, Wednesday, or Friday? If yes, go to friend's house after school; if no, go home.

Activity: A Five-Kingdom Game!

Have students add new creatures to the What Organism Do You See? game and change the flow-chart so that these creatures can be guessed too. They might try making a key that separates the members of the four kingdoms that they have studied: bacteria, protists, plants, and animals. Students could do research and then add the fifth kingdom: fungi!

Students could also do a "treasure hunt" to get specimens from all five kingdoms (or pictures of them) to use in setting up a real key.

Mission 9

What Organism Do You See?

Figure 9.3—Green Seaweed—Unlike other plants that use sap to get nutrients to their cells, green seaweed takes its nutrients directly from seawater.

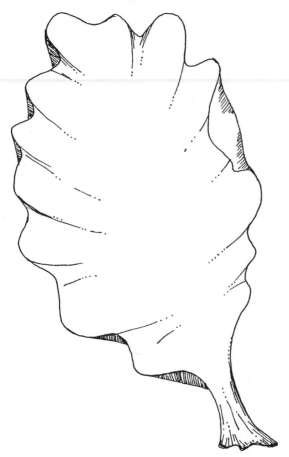

Figure 9.4—Pine Tree—Pine trees are plants whose cells receive their nutrients from sap. Pine trees *do not* produce flowers. They produce cones that contain seeds.

Figure 9.5—Fern—Like most plants, fern cells receive their nutrients from sap. Ferns *do not* produce flowers. They produce spores that are like seeds, but can survive extreme conditions.

Mission 9

What Organism Do You See?

Figure 9.6—Frog—Frogs are animals that have backbones. They are cold-blooded, which means they take on the temperature of their surroundings. Adult frogs have four appendages.

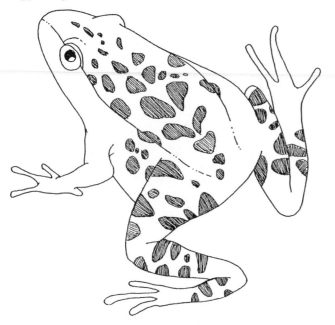

Figure 9.7—Cactus—Like most plants, cactus cells receive their nutrients from sap. Cacti produce flowers. They use their entire structure to produce food in a process called *photosynthesis.*

Figure 9.8—Rosebush—Like most plants, rosebush cells receive their nutrients from sap. Rosebushes produce flowers called *roses.* They use their leaves to produce food in a process called *photosynthesis.*

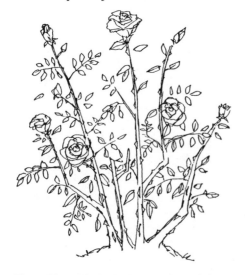

Mission 9

What Organism Do You See?

Figure 9.9—Worm—Earthworms are animals that do not have backbones. Their bodies are divided into segments. Worms have no jointed appendages at all.

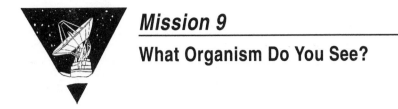

Figure 9.10—Slug—Slugs are animals that do not have backbones and are not composed of segments. Many kinds of slugs are found on land.

Figure 9.11—Scorpion—Scorpions are animals that do not have backbones. Their bodies are divided into segments. Scorpions have 10 appendages, including eight legs and two claws.

Figure 9.12—Centipede—Centipedes are animals that do not have backbones. Their bodies are divided into 12 to 34 segments. Each segment has two appendages.

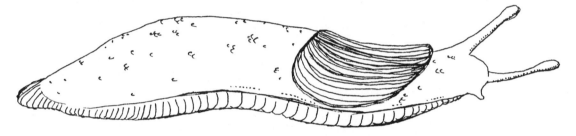

From *How Might Life Evolve on Other Worlds?* © 1995. Teacher Ideas Press. (800) 237-6124.

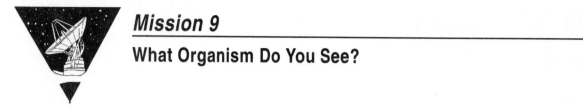

Mission 9

What Organism Do You See?

Figure 9.13—Octopus—Octopuses are animals that do not have backbones and are not composed of segments. They are found in the sea and have eight appendages called *tentacles*.

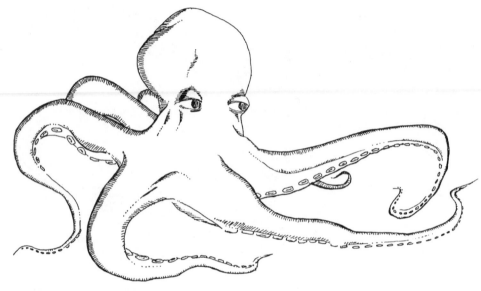

Figure 9.14—Sea Star—Sea stars are animals that do not have backbones and are not composed of segments. They are found in the sea. Most have five appendages called *arms*.

Figure 9.15—Anemone—Anemones are animals that do not have backbones and are not composed of segments. They are found along the seashore, anchored to rocks. They have many appendages called *tentacles*.

Mission 9

What Organism Do You See?

Figure 9.16—Snake—Snakes are animals that have backbones. They are cold-blooded, which means they take on the temperature of their surroundings. They have no appendages.

Figure 9.17—Bear—Bears are animals that have backbones. They are warm-blooded, which means they maintain an even body temperature. Bears are covered with fur.

Figure 9.18—Bird—Birds are animals that have backbones. They are warm-blooded, which means they maintain an even body temperature. Birds are covered with feathers.

From *How Might Life Evolve on Other Worlds?* © 1995. Teacher Ideas Press. (800) 237-6124.

Mission 9

What Organism Do You See?

Mission Briefing

Name:

Date:

Carole Hickman, Biologist on the SETI Academy Team.

One of the most important tasks you will be asked to do at SETI Academy is to imagine the kinds of living creatures, or *organisms*, that are likely to exist on other planets in the galaxy. Again, our starting point in finding out what kind of life may be possible is the only planet with living organisms that we know of—Earth. A fun way to find out about some of the characteristics of Earth life is to play the game What Organism Do You See?

What Do You Think?

1. If you had to sort all the plants and animals on Earth into groups, what characteristics of plants and animals would you look for?

2. What kinds of groups would you use?

Mission 9

What Organism Do You See?

Figure 9.19—Organism Classification Key.

START HERE

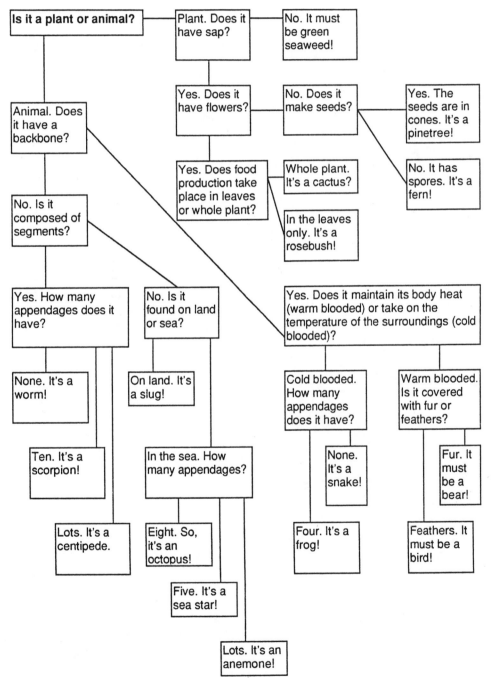

From *How Might Life Evolve on Other Worlds?* © 1995. Teacher Ideas Press. (800) 237-6124.

Mission 9

What Organism Do You See?

What Do You Think, Now?

Name:

Date:

After completing this mission, please answer the following questions:

1. After playing the game, list as many different characteristics of plants or animals on Earth that you can think of.

2. How might characteristics be different for life-forms on other planets?

Mission 10

Inventing Life-Forms
The Creation of an Extraterrestrial Species

Overview

In mission 10, students play a dice-rolling game, Inventing Life-Forms, to simulate the creation of extraterrestrial life-forms. Then they will make illustrations of their life-forms.

Students will consider many of the important characteristics that distinguish different forms of animal life on Earth, such as kind of skeleton, body size and shape, how it gets around, senses, method of reproduction, relationship to other forms of life, and so on, in an evolutionary context. All these characteristics are the result of natural selection; they are all adaptations to environments. Students will also have an opportunity to be creative as they draw pictures and describe the creatures they have invented. As they play the game, students learn that scientists expect biologic concepts to apply to all life, no matter where in the universe it may have evolved.

Concepts

- Scientists expect certain biologic concepts to apply to all life, everywhere in the universe.

- Natural selection operating in an extraterrestrial environment will produce extraterrestrial creatures that are adapted to that environment.

Skills

- Synthesizing knowledge and creative imagination to visualize the evolution of life.

- Follow directions written as a flowchart.

- Understanding the role of chance in events.

Notes

In mission 9, students used a dichotomous classification key to determine the identity of several Earth plants and animals.

153

Mission 10

Materials

For Each Team

- Crayons or markers

- 1 die

For Each Student

- SETI Academy Cadet Logbook

- Pencil

Getting Ready

Before the Lesson

1. (optional) Create reusable "game rules" booklets in folders (the instructions for Inventing Life-Forms are eight pages long).

Classroom Action

1. **Mission Briefing.** Have the class refer to the "Mission Briefing" for mission 10 in their student logbooks while one student reads it aloud.

2. **What Do You Think?** Have students answer the pre-activity questions on the "Mission Briefing." Invite them to share their answers in a class discussion.

3. **Lecture**. Emphasize that scientists who study the possibility of life on other planets have no actual extraterrestrial creatures to study, because none have been discovered yet. However, they can do simulations based on what they know of Earth life and how it evolved. For example, when the supercontinent Pangaea separated into the modern continents, each continent became a "little world" on which evolution proceeded independently: In Australia, marsupials evolved, while in North America, placental ("true") mammals evolved.

    Biologists also know that any organism on another planet must have solved the same problems that all Earth organisms have, although there are many possible solutions to each problem. For example, any organism on any planet must have body openings so that solids and gasses can pass into and out of it to allow for life processes and growth. Also, any organism on any planet must have a method of getting nutrients into its body; the actual method depends on the organism's size. Also, each animal must fit into some sort of food web as both predator and prey. The number of offspring is probably related to the degree that the parent animal cares for its young.

4. **Demonstration**. In the game, Inventing Life-Forms, each student will use a combination of chance probabilities and creativity to imagine one extraterrestrial (Planet Y) creature. Explain to students the rules of this game:

   a. Each student rolls one die to find out one characteristic of their extraterrestrial creature, carefully following the instructions on pages 160-66.

   b. Each characteristic is written on the "Recording Sheet" in the logbook along with the name of an Earth creature that has that same characteristic. Explain that the purpose of listing the Earth creature is to help visualize how that characteristic affects the life and behavior of the Earth animal. The extraterrestrial animal may be similar!

   c. The student proceeds to the next step, following the directions very closely. There are a few cases where they will have to roll the die twice, or skip a step.

   As an example, roll a die for step 1—"Skin"—and show how to record the results of the roll. Point out the direction "Go to" that tells what step to do next.

   Ask students to look at step 24 and the "My Extraterrestrial Species" logbook sheet, where they name, draw a picture of, and describe their creature. Students who finish early can ask for drawing materials and go on to that step. Teachers may wish to make completion of the last step a homework assignment.

Teacher's Note: Students will want to go from step to step following the numbers rather than the directions. Watch for this when you move around the room.

5. **Activity**. Give students the remainder of the class period to complete the game. Help as needed, and check to see that all students are recording their results. Small drawings of plant and animal forms used for food can be done on the back of the recording sheet when students draw their life-forms, or on a separate sheet of paper.

Students should have fun with this game and recognize that life on other worlds will follow the rules that have led to the development of life on Earth, though the outcomes will be different. This is like dealing a hand of cards. Following the rules of the game, a specific number of cards will be dealt. However, each actual hand is a different outcome.

Closure

1. **Discussion**. Post students' drawings on a wall. Allow time for each student to describe their creature. Ask students to tell how intelligent they think their creature might be, but do not emphasize intelligence as it is a complex topic, and the subject of *The Rise of Intelligence and Culture* in the *SETI Academy Planet Project* volumes.

2. **What Do You Think, Now?** Have students answer the post-activity questions on the logbook sheet "What Do You Think, Now?" Invite students to share their responses. Ask students how their opinions have been changed by this mission. In response to these questions, students might recognize, for example, that many different life-forms could exist in different habitats on the same planet. They might also see that characteristics implying the amount of gravity or composition of the atmosphere might mean that the creatures must be from different planets. For example, on a planet with a thin atmosphere, land-dwelling creatures might have huge lungs. On a massive planet with strong gravity, creatures might be flat and low to the ground. Encourage open discussion and alternative points of view.

Going Further

Activity: Creative Writing

Invite students to make up a story from the viewpoint of their creature. They might want to describe a day in the life of their creature, or what happens when their creatures meet a visitor from Earth.

Activity: Encore, More Extraterrestrials!

Have students invent another creature from the same planet, using only their imagination (no die-rolling). They should describe all the characteristics of their creatures and how their new invented animals or plants are related to the creature they created with die-rolling in the game.

Mission 10

Inventing Life-Forms

Mission Briefing

Name:

Date:

Dave Milne, Biologist on the SETI Academy Team.

All living things took their present form through a very long series of chance variations. Most of the changed creatures died, but the few that were better adapted to their environment survived, passing on those changes to their offspring. This is natural selection. In this game, you are invited to create a form of animal life that might have evolved on a planet that is somewhat, but not exactly like, Earth. Rolling a die will simulate the role of chance in the evolution of your extraterrestrial creature. Each roll of the die represents a result of millions of years of chance variations that were successful in the extraterrestrial environment. At the end of the game, you will have a chance to be creative and draw a picture showing what your extraterrestrial looks like.

What Do You Think?

1. In what ways might a habitable extraterrestrial planet be different from Earth?

2. In what ways could extraterrestrial life-forms be different from Earth life-forms?

Mission 10

Inventing Life-Forms

Name:

Date:

Table 10.1—Student Recording Sheet.

	Adaptation	Adaptation of Your Extraterrestrial Organism	Earth Animal with Adaptation
1.	Kind of skin		
2.	Number of openings		
3.	Long or other		
4.	Segments		
5.	No. Appendages/ No. Segments		
6.	Hard parts		
7.	Hard outside parts		
8.	Size		
9.	Feeding cells		
10.	Moving around		
11.	Sensing vibrations		
12.	Chemical senses		
13.	Number of eyes		
14.	Eating		
15.	Plant-eater		
16.	Predator		
17.	Defensive structures		
18.	Poison as a defense		
19.	Defensive behaviors		
20.	Reproduction		
21.	Sexual reproduction		
22.	Mating		
23.	Babies		

From *How Might Life Evolve on Other Worlds?* © 1995. Teacher Ideas Press. (800) 237-6124.

Mission 10

Inventing Life-Forms

Directions

On each roll, use one die. Record all your adaptations into the center column of the "Recording Sheet" in your logbook, and copy the Earth animal (in parentheses) with the same attribute in the column on the right.

---

1. Skin
If you roll a . . .
1, 2, or 3 = you have scaly skin (*like a snake*).
4 or 5 = you have mucus-covered soft skin (*like a slug*).
6 = you have leathery skin (*like a cow*).
Record this result on your recording sheet.
Go to—**2. Openings**

---

2. Openings
    People have different openings or holes in their bodies to breathe, get rid of waste, and eat. Some animals, like anemones, use one opening for several of these purposes. How many openings does your organism have, and which does it use for what?
If you roll a . . .
1 or 2 = 1 opening (*like a sea anemone*).
3 or 4 = 2 openings (*like an earthworm*).
5 or 6 = 3 openings (*like a frog*).
Record this result on your recording sheet.
Go to—**3. Body Shape**

---

3. Body Shape
If you roll a . . .
1, 2, 3, or 4 = your organism's body is longer than it is wide (*like a worm*).
5 or 6 = your organism has some other shape.
Record this result on your recording sheet.
Go to—**4. Segments**

---

4. Segments
If you roll a . . .
1, 2 , or 3 = your organism's body is divided into segments (*like a centipede*).
Record this result on your recording sheet.
Go to—**5a. Appendages for Segments**
4, 5, or 6 = your organism's body has no segments (*like a toad*).
Record this result on your recording sheet.
Go to—**5b. Appendages for No Segments**

5a. Appendages for Segments

Roll one die two times and add the numbers together to find the number of segments your organism has. Your organism has two appendages on every segment (example: 3 segments x 2 = 6 appendages).

Record this result on your recording sheet.

Go to—**6. Hard Parts**

5b. Appendages for No Segments

People have four appendages: two arms and two legs. On Earth, most animals have an even number of appendages, but that may not be the case on all planets. Roll one die. This is the number of appendages your organism has.

Record this result on your recording sheet.

Go to—**6. Hard Parts**

6. Hard Parts

If you roll a . . .

1, 2, or 3 = your organism has hard parts on the outside of its body (*like a lobster*).

Record this result on your recording sheet.

Go to—**7. Outside Hard Parts**

4 or 5 = your organism has hard parts on the inside of its body (*bones, like you*).

Record this result on your recording sheet.

Go to—**8b. Large Sizes**

6 = your organism has no hard parts. It gets around by wiggling (*like a worm*).

Record this result on your recording sheet.

Go to—**8a. Small Sizes**

7. Outside Hard Parts

If you roll a . . .

1 or 2 = your organism has a hard shell (*like a snail*).

Record this result on your recording sheet.

Go to—**8a. Small Sizes**

3 or 4 = your organism has a protein, armor-like covering (*like a beetle*). It must shed its armored skin to grow.

Record this result on your recording sheet.

Go to—**8a. Small Sizes**

5 or 6 = your organism has a protein, shell-like covering (*like an insect*). It doesn't need to shed to grow.

Record this result on your recording sheet.

Go to—**8b. Large Sizes**

8a. Small Sizes

Roll one die for the range of your organism's weight.

If you roll a . . .

1 or 2 = less than 1 pound (*like a mouse*).

3 or 4 = 1-2 pounds (*like a rat*).

5 or 6 = 2-5 pounds (*like a chicken*).

Record this result on your recording sheet.

Go to—**9a. Feeding the Cells of Small Animals**

8b. Large Sizes

Roll one die for the range of your organism's weight.

If you roll a . . .

1 = 6-9 pounds (*like a cat*).

2 = 10-49 pounds (*like a bobcat*).

3 = 50-99 pounds (*like a German shepherd*).

4 = 100-199 pounds (*like an alligator*).

5 = 200-999 pounds (*like a pig*).

6 = 1,000-100,000 pounds (*like a dinosaur*).

Record this result on your recording sheet.

Go to—**9b. Feeding the Cells of Large Animals**

9a. Feeding the Cells of Small Animals

Like people, earthworms take in oxygen and food from their environment. The oxygen and food nutrients are absorbed by the worm's blood. Five small hearts pump the blood through the body to give oxygen and food nutrients to all the cells in the worm's body. Other small animals get food and oxygen to their cells in different ways. Roll one die to find out how your small animal gets oxygen and nutrients to its cells.

If you roll a . . .

1 or 2 = oxygen and food are in the blood, which is pumped to organs by one pump (heart) (*like a mouse*).

3 or 4 = oxygen and food are in the blood, which is pumped to organs by more than one pump (several hearts: Roll one die for the number of hearts; if you roll a one, roll again until you get a larger number) (*like an earthworm*).

5 = oxygen and food are in the blood, which sloshes around inside your animal, bathing all cells (*like a lobster*).

6 = your organism is only a few cells thick. It absorbs oxygen and food through its skin and has no blood (*like a flatworm*).

Record this result on your recording sheet.

Go to—**10. Moving Around**

9b. Feeding the Cells of Large Animals

We take in air and food from our environment. Our blood absorbs oxygen from the air and nutrients from the food. Our hearts pump the blood through our bodies, carrying the oxygen and nutrients to every cell of our bodies. Roll one die to find how your animal gets oxygen and nutrients to its organs.

If you roll a . . .

1, 2, 3, or 4 = oxygen and food are in the blood, which is pumped to organs by one pump (heart) (*like you*).

5 or 6 = oxygen and food are in the blood, which is pumped to organs by more than one pump (several hearts: Roll one die for the number of hearts) (*like an octopus*).

Record this result on your recording sheet.

Go to—**10. Moving Around**

10. Moving Around

If you roll a . . .

1 = crawls on land (*like a snail*).

2 = walks on land (*like a centipede, beetle, lizard, or ostrich*).

3 = swims in water (*like a fish*).

4 = drifts in water (*like a jellyfish*).

5 = jet propulsion in water (shoots out a burst of water through one of the holes in its back end, propelling the organism forward) (*like a squid*).

6 = flies in atmosphere (if larger than 100 pounds, roll again) (*like a bird or bat*).

Record this result on your recording sheet.

Go to—**11. Sensing Vibrations**

11. Sensing Vibrations

People use their ears and sense of touch to feel vibrations in the air and the ground. What does your organism use to sense vibrations?

If you roll a . . .

1, 2, or 3 = your organism has organs (like ears) to sense vibrations in air (*like the whiskers on a mole*) or in water (*like you*).

4, 5, or 6 = your organism uses small hairs scattered over its body to sense vibrations in the air or water (*like some fishes or hairy tarantulas*).

Record this result on your recording sheet.

Go to—**12. Chemical Senses**

12. Chemical Senses

People have sensors that sense chemicals in the air, food, or water. They are in your nose and tongue, but chemical sensors are not always in noses or on tongues. Spiders have them on the soles of their feet. Where are your organisms chemical sensors located?

If you roll a . . .

1, 2, or 3 = one place (*like a spider*).

4, 5, or 6 = two places (*like you*).

Record this result on your recording sheet.

Go to—**13. Sensing Light**

---

13. Number of Eyes
If you roll a . . .
1 or 2 = no eyes (*like blind cavefish*).
3 or 4 = two eyes (*like you*).
5 or 6 = more than two eyes (*like a spider*).
Record this result on your recording sheet.
Go to—**14. Eating**

---

14. Eating
If you roll a . . .
1, 2, 3 , or 4 = plant-eater. Needs something with which to snip and grind plant
    parts (*like a cow*).
Record this result on your recording sheet.
Go to—**15. Plant-Eater**
5 = meat-eater. Needs to have meat tearers (like claws and sharp teeth) (*like a
    tiger*).
Record this result on your recording sheet.
Go to—**16. Predator**
6 = plant-eater and meat-eater. Needs to have both plant grinders and meat
    tearers (*like a bear*).
Record this result on your recording sheet.
Go to—**16. Predator**

---

15. Plant-Eater
Your organism must protect itself against predators.
  Go to—**17. Protection—Defensive Structures**

---

16. Predator
    How does your organism catch its prey? Roll one die twice and add the
numbers to get one adaptation. Repeat to get a second adaptation.
If you roll a . . .
2 = chase it (*like a cheetah*).
3 = hit it (*like a hawk*).
4 = suffocate it (*like a python*).
5 = blind it(*like a spitting cobra*).
6 = spear it (*like a harpoon worm*).
7 = inject poison (*like a scorpion*).
8 = make a trap (*like a spider*).
9 = lure it (trick it with a treat) (*like an angler fish*).
10 = electric shock (*like an electric eel*).
11 = stun it with vibrations (*like a dolphin*).
12 = work together with others of the same species (*like wolves*).
Record this result on your recording sheet.
  Go to—**17. Protection—Defensive Structures**

17. Protection—Defensive Structures

If you roll a . . .

1 or 2 = spines (*like a porcupine*).

3 or 4 = thick, protective covering (*like a turtle*).

5 or 6 = horns (*like a triceratops*).

Record this result on your recording sheet.

Go to—**18. Protection—Poison as a Defense**

18. Protection—Poison as a Defense

If you roll a . . .

1, 2, 3, 4, or 5 = no, your organism is not poisonous. It has camouflaged coloring, so it blends in with the planet's most common plant color (*like a grasshopper*).

6 = yes, your organism is poisonous to eat, or has a venomous sting, and has coloring that warns other animals to leave it alone by standing out against the planet's plant color (*like poison arrow frogs*).

Record this result on your recording sheet.

Go to—**19. Protection—Defensive Behaviors**

19. Protection—Defensive Behaviors

If you roll a . . .

1 = run away (*like an antelope*).

2 = hide (*like a prairie dog*).

3 = freeze (stand very still) (*like a pheasant*).

4 = fight (*like cats and dogs*).

5 = pretend you're bigger than you really are (*like a scared cat*).

6 = work together with others of your species (*like a musk ox*).

Record this result on your recording sheet.

Go to—**20. Reproduction**

20. Reproduction

If you roll a . . .

1, 2, or 3 = asexual: Your organism reproduces without sex by budding off a piece of itself, which grows into another identical organism (*like a sea anemone*).

Record this result on your recording sheet.

Go to—**24. Design Your Organism**

4, 5, or 6 = sexual: Your organism reproduces by sex (two organisms share their adaptations, and make babies a little different from themselves) (*like cats*).

Record this result on your recording sheet.

Go to—**21. Sexual Reproduction**

21. Sexual Reproduction

If you roll a . . .

1 or 2 = your species has two sexes: A and B. Every individual has only one sex (*like people and other mammals*).

3 or 4 = your species has two sexes: A and B. Every individual has both sexes, but two individuals are required to mate (*like earthworms*).

5 or 6 = your species has three sexes: A, B, and C. Every individual has one sex. Three individuals are required to mate (*like no examples on Earth, but this is a Planet Y organism*).

Record this result on your recording sheet.

Go to—**22. Mating**

22. Mating

Because your species has to get together all its sexes to mate, they need a way to find each other.

If you roll a . . .

1 or 2 = mating call (*like a song sparrow*).

3 or 4 = bright colors (*like a male pheasant*).

5 = mating dance (*like a male peacock*).

6 = mating smell (*like big moths*).

Record this result on your recording sheet.

Go to—**23. Babies**

23. Babies

If you roll a . . .

1 = lay many eggs, but leave and don't take care of the babies (*like fish*).

2 = lay few eggs and take care of eggs and young. Whichever sex takes care of the eggs and young must be camouflaged (*like birds*).

3 or 4 = live birth, care for young in some form of nest. Whichever sex(es) take care of the eggs, the young must be camouflaged (*like rabbits*).

5 or 6 = live birth, care for young in a pouch (*like a kangaroo*).

Record this result on your recording sheet.

Go to—**24. Design Your Organism**

24. Design Your Organism

Using the attributes you now have for your extraterrestrial organism:

1. Draw what your organism and its extraterrestrial food source look like. Give the species a name.

2. Have a friend check your drawing to see that it has the attributes on your recording sheet.

3. Write a page about your creature, telling about all the attributes not shown in the drawing, plus any other information you would like to add.

Mission 10

Inventing Life-Forms
My Extraterrestrial Species

Name:

Date:

Name of Species:_____

Appearance of Species

Food for Species

Mission 10

Inventing Life-Forms

What Do You Think, Now?

Name:

Date:

After you have completed this mission, please answer the following questions:

1. In what ways do you think the planet where your extraterrestrial creature lives might be different from Earth?

2. Compare your extraterrestrial creature with other students' creatures. Could they all survive on the same planet? Why or why not?

Mission 11

Creating Your Extraterrestrial's Family Tree
How Did Your Extraterrestrial Evolve?

Overview

By conducting simulations such as Inventing Life-Forms from mission 10, SETI biologists can imagine a few of the countless kinds of life-forms that might actually exist in the universe.

In mission 11, students invent reasonable evolutionary family trees for their extraterrestrial life-forms. Students will also have an opportunity to be creative as they draw a picture and describe the evolution of the creature they have invented. In explaining their creatures, the students should imagine what preexisting features gave rise to the features that their organisms now have, based on the idea of *descent with modification*. If students have completed the missions in *The Evolution of a Planetary System* of the *SETI Academy Planet Project*, then tell them that they are describing the evolution of inhabitants of Planet Y.

Concepts

- If life does exist on other planets, the life-forms themselves will probably be different from those on Earth, but the process of evolution will be similar to that process on Earth.

- Natural selection operating in an extraterrestrial environment will produce extraterrestrial creatures that are adapted to that environment.

- Evolution often proceeds from simpler to more complex and varied life-forms.

- Evolution on Earth often took hundreds of millions of years to produce different species; it is likely that evolution on extraterrestrial worlds will also require long periods of time.

- Every structure evolves from some other structure. This is called descent with modification.

Figure 11.1—Extraterrestrial Ancestors.

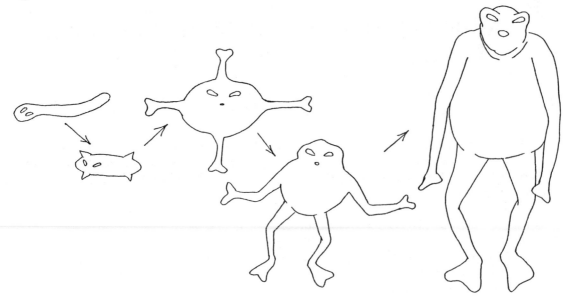

Skills

- Arranging evidence.

- Using a simple key.

Mission 11

Materials

For the Class

- Biologic timeline of life on Earth from mission 7

- Example of a family tree from mission 8

For Each Team or Pair

- Colored markers

For Each Student

- SETI Academy Cadet Logbook

- Sheet of drawing paper, 11 by 14 inches or larger

- Scratch paper

- Extraterrestrial life-form from mission 10

- Pencil

Getting Ready

Just Before the Lesson

1. Post the family tree and the timeline (from mission 8) where everyone can see them.

2. On the chalkboard, draw a simple example, such as figure 11.1 (see page 170), which is one possible evolutionary sequence, showing how the worm-like creature shown on their instruction sheet might evolve.

Classroom Action

1. **Mission Briefing**. Have the class refer to the "Mission Briefing" for mission 11 in their student logbooks while one student reads it aloud.

2. **What Do You Think?** Have students answer the pre-activity questions on the "Mission Briefing." Invite them to share their answers in a class discussion.

3. **Discussion**. Direct students' attention to the family tree that is displayed. Ask students to recall some of the things they learned about evolution during mission 8. Remind them that they used similarities in structures to trace possible lines of descent. As one species evolves into another species, one structure may evolve into another structure.

   Ask students to brainstorm some of the features of an animal that might change over millions of years. Write these on the chalkboard. For example, the list might include:

   - use and length of appendages
   - senses such as eyes, ears, and nose
   - body shape and size
   - position of features
   - color and texture of skin
   - ability to survive on different kinds of foods
   - change of habitat—land to water or water to land

   Relate items from the list to the evolution of Earth animals on the family tree. For example, a worm-like animal evolved into a fish as its basic body shape remained similar (long), while, over these great expanses of time, it developed an internal *notochord*, which eventually became the bony spinal column. It also evolved fins from muscles and skin flaps. Bones that moved the fins formed; this contributed to effective locomotion in the water. Scales developed, which made for a more streamlined body. Point out that the lobe-finned fish gave rise to both amphibians and reptiles. One species can branch into two or more new species. Demonstrate this on the chalkboard: Draw a simple family tree and then add another branch, showing how two different species might have evolved from an earlier form.

4. **Demonstration**. Explain that each student is supposed to create their own family tree showing how a worm-like creature might have evolved into the extraterrestrial life-form they created during mission 10. Students must find the intermediate stages in its evolution from a simple early ancestor. This ancestor may be thought of as a simple "worm-like" organism, or even a simple bacterial cell. Students in each team should sketch their family trees on scratch paper first before making a final drawing.

Select one student's creature to use as an example. Identify two or three prominent features. Prompt students to think about what those features might be descended from in the creature's ancestor (only a few generations back). (Encourage anything even vaguely plausible. The idea at this point is to encourage students to think again about the fact that structures evolve from preexisting structures, not to know in detail the ways in which they did so on Earth.) Then encourage students to draw this ancestral creature. Use the same approach each time an even earlier (more worm-like) ancestor is envisioned.

Table 11.1 gives examples of features of Earth organisms and the structures from which they evolved (this material is repeated from mission 8). These examples illustrate that the structures of organisms are traceable to earlier structures in their ancestors. Use this information to evaluate students' evolution of their extraterrestrial creatures. Copy and distribute this list as a spur to the students' imaginations.

Table 11.1—Features of Earth Organisms.

Feature	Structure from Which It Evolved
Eyes	Simple, light-sensitive cell, then lots side by side
Limbs	In vertebrates, from fins on fish-like ancestors In insects and centipedes, from flaps on side walls of worm-like ancestors
Wings	In vertebrates, from forelimbs of reptiles or mammals In insects, from stiff flaps extended from body wall
Poison fangs	In snakes, grooved fangs became hollow fangs; salivary glands secrete toxin In centipedes, spiders, "fangs" are stiff, sharp front limbs connected to salivary glands
Armor	In turtles, expanded rib bones fused to hardened skin In armadillos, expanded hard lumps in skin form armor
Horns	Growths of bone from skull covered by hardened skin (antelopes, cattle), groups of hard, stiff hairs (rhinos), bony growths after skin falls off (deer); hardened exoskeleton (beetles); shell secreted by glands in skin (molluscs)
Stings	Hardened, sharp ovipositor in neuter (female) worker bees and wasps; hard exoskeleton spine (scorpions)
Hooves	Enlarged, hardened toenails
Tusks	Elongated teeth (elephant, saber-tooths, walruses)
Feathers, hair	Modified reptilian scales
Electric shock ability	In electric eel; exaggeration of ability of some fish to use electric fields for navigation; electricity is generated by activities of many cells (for example your heart, as seen by EKG); these fish have enlarged arrays of these cells
Lungs	In vertebrates, modified swim bladder of fish ancestor (which was a gas-filled space used for buoyancy in water) In land snails/slugs, the space that once held the gill of the aquatic ancestor, now lined with lots of blood vessels
Eardrums	Modified small bones that once worked the jaw joint in fish; in insects a modified plate of the exoskeleton
Jaws	In vertebrates, the bent, enlarged, bony joints that once supported a gill In insects, centipedes, spiders, crustaceans, mouthparts are modified front limbs that were once on the front end of a more centipede-like ancestor

5. **Activity**. Form teams of two or three students each. Distribute the drawing materials and have students draw several earlier stages in the evolution of their invented organism. As students in each team finish sketching their family trees, have them make finished drawings.

   (optional) Create a timeline for these stages.

Closure

1. **Discussion**. Ask for a few volunteers to explain their family trees to the entire class. Point out the timeline for the biologic evolution of life on Earth. Ask students how long it took for life to evolve from one-celled organisms to multicelled but simple animals like worms (about three billion years). How long did it take for the complex life-forms we know today to evolve from simple multicelled life-forms like worms (hundreds of millions of years). Tell students that biologists believe that evolution from simpler to more complex species would take roughly the same amounts of time on other worlds as it takes on Earth.

2. **What Do You Think, Now?** Have students answer the post-activity questions on the logbook sheet "What Do You Think, Now?" Invite students to share their responses. Ask students how their opinions have been changed by this mission.

Going Further

The Next Book with SETI Academy

In *The Rise of Intelligence and Culture* of the *SETI Academy Planet Project*, students explore how extraterrestrial organisms might develop intelligence and cultures on other worlds. Teachers not planning to go on to *The Rise of Intelligence and Culture* should pose these open-ended challenges for students as shown below.

Activity: Where Does Your Extraterrestrial Live?

Have students create a map of the planet on which their extraterrestrials live. (Or use the map of Planet Y if students have completed the missions in *The Evolution of a Planetary System* of the *SETI Academy Planet Project*.) Show the continents, oceans, and areas where extraterrestrial creatures can be found.

Activity: Create an Environment

Have students create an environment for the planet above, describing how plants and animals interact. Are any of the species intelligent? If so, which are more intelligent than others? Do they live in peace or do they fight? What are their homes like? Do they live in groups or alone?

Discussion: Sharing Ideas

Have students describe changes to the environment or to the planet that might have contributed to the evolution of their extraterrestrial life-forms. Have students create a scenario for each node of their extraterrestrial's family tree.

Activity: Humans Travel to Planet Y!

If students have completed the missions in *The Evolution of a Planetary System* of the *SETI Academy Planet Project*, have them write a story describing what happens when people from Earth visit Planet Y.

Have students suggest things that might be different about Planet Y and its star that might affect the rate at which evolution takes place. Some ideas the students may have are:

- greater seasonal variability in temperatures

- length of the day and year

- x-ray or ultraviolet flux from the star

- rate of asteroid impacts

- degree of volcanic activity

- strength of plate tectonics

- original atmosphere composition

- other changes in atmosphere (loss or buildup of CO_2).

Mission 11

Creating Your Extraterrestrial's Family Tree
Mission Briefing

Name:

Date:

John Oro, Biochemist on the SETI Academy Team.

Just as every organism on Earth has evolved through a series of intermediate stages, your extraterrestrial life-form must also have evolutionary ancestors. In this mission, you are asked to imagine the evolutionary family tree that may have led to the evolution of your extraterrestrial life-form.

What Do You Think?

1. How would a family tree of your extraterrestrial life-form be the same as a family tree of Earth animals?

2. What might the differences be?

Mission 11

Creating Your Extraterrestrial's Family Tree

Name:

Date:

1. Assume that a planaria-like worm was the oldest *complex* ancestor of your extraterrestrial life-form. (The oldest *simple* ancestors were probably similar to the single-celled organisms like cyanobacteria and protists.)

Figure 11.2—The Original Worm!

2. Draw the evolutionary descent of your life-form. On scratch paper, draw five steps of the family tree. The most recent step is the life-form from the last mission, and the first step is the worm-like organism. The in-between steps are up to you, but they need to be scientifically realistic.

Figure 11.3—Evolution of Your Life-Form.

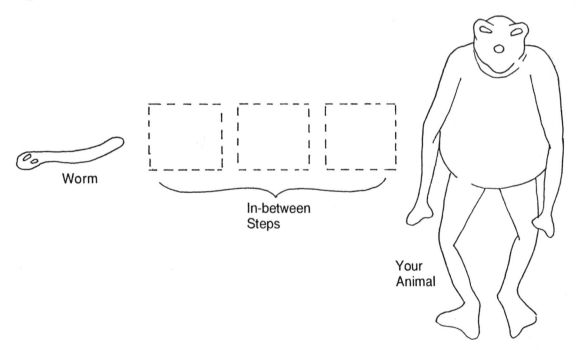

3. Imagine some organisms that might have evolved from the three in-between-step creatures and create more branches for your family tree.

4. Check your drawings and make any changes or corrections. Now draw the final draft of your extraterrestrial's family tree on the drawing paper. Use colored markers so it can be seen from a distance.

Mission 11

Creating Your Extraterrestrial's Family Tree

What Do You Think, Now?

Name:

Date:

After you have completed this mission, please answer the following questions:

1. Compare the family tree of your extraterrestrial life-form with the family tree of Earth animals. How are they similar or different?

2. Choose one of the creatures from one of the in-between steps. Imagine that it evolved into another organism, like a certain dinosaur that evolved into both a large dinosaur and a small bird about 250 to 150 million years ago. Draw a few creatures that evolved from that creature and add branches to your extraterrestrial's family tree.

Mission 12

Mission Completed! What Have You Learned?

Overview

In this short session, students write and draw what they have learned about the evolution of simple and complex life on Earth and the time involved in evolution. This session may be used as an assessment of what students have learned.

Mission 12

Materials

For Each Student

• SETI Academy Cadet Logbook

• Pencil

Getting Ready

No preparation is necessary.

Classroom Action

1. **Mission Briefing**. Have the class refer to the "Mission Briefing" for mission 12 in their student logbooks while one student reads it aloud.

2. **Activity**. Have students review the questions in the "Mission Briefing." Ask them to answer the questions in full and complete sentences, and in as much detail as possible, using extra paper if necessary. Give the students 15 to 30 minutes for this.

Closure

1. **Discussion**. Invite students to share their drawings, ideas, and answers. Correct any misconceptions, or suggest alternatives.

Achievement Award

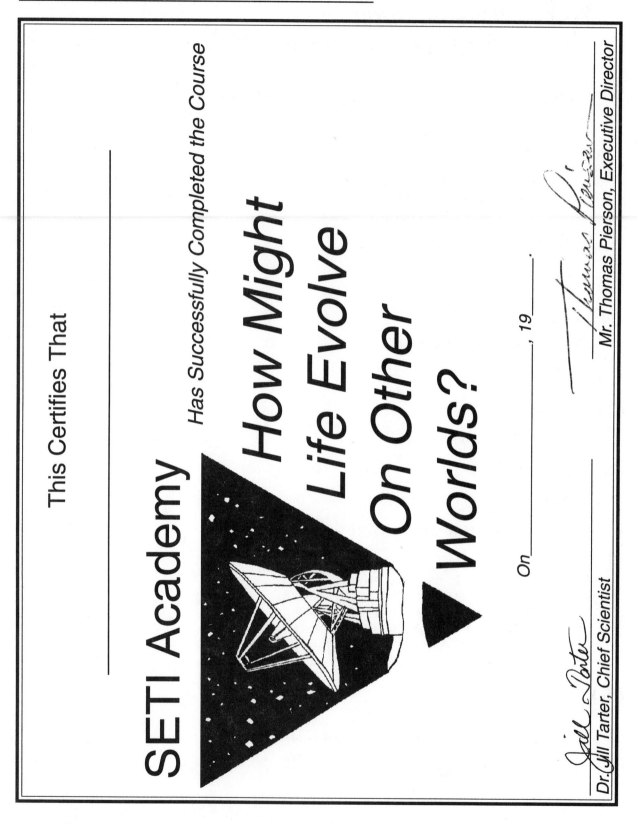

This Certifies That

_____

Has Successfully Completed the Course

SETI Academy

How Might Life Evolve On Other Worlds?

On _____, 19____.

Dr. Jill Tarter, Chief Scientist

Mr. Thomas Pierson, Executive Director

Mission 12

Mission Completed!

Mission Briefing

Name:

Date:

Jill Tarter, Chief Project Scientist at the SETI Institute.

Congratulations on completing your missions at SETI Academy! You now have a good background in what SETI scientists do when they consider the forms of life that they might find on other planets. In the future, we hope that you will enjoy thinking about the possibilities of life on other worlds, and that someday you might even want to join us on the staff at the SETI Institute.

In this last mission, we would like you to think back over our course of study and give your best answers to questions on these three pages. Use additional paper if necessary to answer as fully as possible.

1. As you have seen, for most of our planet's lifetime, visitors from space would only have detected microscopic life-forms. Complex life evolved only in the last few hundred million years, and people have been present only for the last tiny fraction of Earth's history. What does this tell us about what we might expect when we look for life on planets around other stars?

From *How Might Life Evolve on Other Worlds?* © 1995. Teacher Ideas Press. (800) 237-6124.

2. On this page, draw a family tree for vertebrates (animals with backbones). Hint: Some common vertebrates existing today are mammals, fish, reptiles, amphibians, and birds. When you create your family tree, put the oldest organism(s) on the left side of the page and the youngest on the right side of the page.

3. Suppose you meet an extraterrestrial who has no idea about the origin and evolution of life. Write a letter to this extraterrestrial explaining what you know about evolution, and what evidence the extraterrestrial might look for to discover its own family tree.

4. You learned in previous activities that Earth was formed about 4.5 billion years ago. On the family tree you drew on the previous page, add your best guess about how long ago the various vertebrate species first appeared.

5. Draw a magnified view of what you think some of the oldest living things on Earth looked like. About how large were they? What were they called?

6. Would you like to continue at SETI Academy to learn about the evolution of intelligent life on other worlds?

7. What especially would you like to learn more about?

# Glossary

Algae, green. Primitive multicellular plants also found in or by water without true stems, roots, or leaves but containing chlorophyll.

Appendages. Anything that is attached to the outside of an animal. Arms and legs are appendages.

Astronomer. A person who studies space and everything in it.

Bacteria. One-celled microorganisms with simple cells that have no chlorophyll; they multiply by simple cell division.

Blue-green bacteria. See "Cyanobacteria."

Biologist. A person who studies all living things, including plants and animals.

BYA. Billion years ago.

Characteristics. Qualities or traits by which an organism is grouped or classified according to a system or category.

Climatologist. A person who studies the climate or weather.

Cyanobacteria. Blue-green bacteria; single-celled organisms with simple cells that make their own food.

Evolution. A gradual change of something into a significantly different form, often more complex or sophisticated.

Exoskeleton. An external structure used for protection and/or support.

Extraterrestrial. Anything not of or from Earth.

Fossil. The preserved remains of an organism; sometimes the bones, sometimes just the impression of the organism.

Geologist. A scientist who studies the history and structure of Earth.

Hominids. A primate of the family Hominidae of which modern man is the only living species.

***Homo sapiens*.** The genus and species for human beings.

Hypothesis. A reasoned guess.

Invertebrate. An animal without a backbone.

Microscopic. Something too small to be seen without a microscope.

MYA. Million years ago.

Natural selection. The way that evolution works. Organisms with better variations live and pass on the good traits to their young.

Paleontologist. A scientist who studies fossils.

Pangaea. The supercontinent that existed 200 million years ago on Earth, when the present continents were united as one.

Photosynthesis. The way a plant makes food from sunlight, water, air, and minerals.

Planaria. Flatworms having broad, flat bodies with microscopic hair-like cells and a three-branched digestive cavity; multicellular animals.

Planetologist. A scientist who studies planets.

Plate tectonics. The movements of large pieces of Earth's crust, which result in mountain building, volcanoes, earthquakes, and changes in the arrangement and positions of the continents.

Protists. A large group of microscopic one-celled animals with complex cells; some make their own food and some need to find their food.

SETI. Search for Extraterrestrial Intelligence.

Stellar astronomer. A scientist who studies stars.

Theory. The best possible scientific explanation for what is observed; supported by much evidence and not contradicted by any evidence. *Evolution* is a theory.

Vertebrate. An animal with a backbone.

Appendixes

Suggestions for Cutting Costs

1. **Options**. Omit some or all of the optional materials. These materials are indicated in the tables that follow.

2. **Teams**. We recommend working with students in teams of two, for more hands-on interaction. However, to save money or materials, have students work in teams of four (or even more). This will cut the cost of many of the needed materials in half (or even more). These materials are indicated in the tables that follow.

3. **Demonstrations**. Do some of the laboratory work as a class demonstration instead of as a hands-on exercise. This is especially true for mission 4.

4. **Home Culture**. The easiest way to get living protists, green algae, and other organisms is to order them from a biological supply company. However, it is cheaper to gather your own from local ponds, rivers, and aquariums. Or, make hay infusions (students will be the first to say that these stink!) and grow bacteria on potato slices.

5. **Substitutions**. Some substitutions can be made. These substitutions are indicated in the tables that follow.

6. **Centers**. It is recommended that each team be given their own materials. In some cases, it is possible to set up a central workstation and use fewer materials. Setup time is longer.

Required Materials List

"Living" Materials

Materials that must be ordered weeks in advance of use are printed in boldface. Provide the date you will be using the live materials when you order from the company.

Table A.1—Living Materials.

Material	Substitutions or Alternatives. Optional Items Are Indicated.	Quantity per Pair, Team, or Center	Quantity for Each Class of 32	Reusable in Each Class	Used in Activity
Planaria, brown	Other flatworms		1 jar	Yes	3, 4
Protists	Pond water		1 jar	Yes	3, 4
Cyanobacteria	Aquarium water		1 jar	Yes	3, 4
Green algae	Aquarium water		1 jar	Yes	3, 4
Bacteria	Potato culture		1 culture	Yes	3, 4
Live animals	Pictures of animals		Varies	Yes	7

Office, Art, and General Supplies

Table A.2—Office Supplies and Art Materials.

Material	Substitutions or Alternatives. Optional Items Are Indicated.	Quantity per Pair, Team, or Center	Quantity for Each Class of 32	Reusable in Each Class	Used in Activity
Drawing materials	Colored marker, pens, pencils, or crayons		Variable	Yes	All
Black construction paper, 9 in. X 12 in.	Another dark color		3	Yes	2
Plastic cm. rulers	Meter sticks	1		Yes	2, 9
Paper towels	Cloth dishtowels		1 roll	No	3
Masking Tape		2-3 pieces	2-3 rolls	No	4,9
Clock with second hand	Watch, stop-watch, wall clock		1	Yess	5, 6
Glass jar	Water glass	1		Maybe	5
Sand	Gravel	Variable	About 1 gallon	Not easily	5
Dirt	Planting mix	Variable	About 1 gallon	Not easily	5
3 kinds of tiny plastic animals or shells	Pictures or stickers of animals	1 (or more) each of 3 kinds		Not easily	5
Water			1 gallon	No	5
Three colors of construction paper	Newspaper (for one "color")	6 sheets of each color		5 sheets yes	5
Fossils	Optional		Variable	Yes	5
Shale, sandstone	Optional		1 or each	Yes	5
Transparent tape		1 roll		No	5, 7, 8
Scissors		1		Yes	5, 7, 8
Chopsticks	Any gripping tool, tweezers.	1 pair		Yes	6
Pliers	Any gripping tool	1 pair		Yes	6
Salad tongs	Any gripping tool	1 pair		Yes	6
Dried beans, peas	Other dried seeds	100 each		yes	6
Small cups	Paper or plastic cups	3		Yes	6
Carpet scrap	Cloth scrap, paper	1		Yes	6
Plastic bag	Paper bag	1		Yes	6
Calculator	Optional	1		Yes	6

Table A.2—*continued*

Material	Substitutions or Alternatives. Optional Items Are Indicated.	Quantity per Pair, Team, or Center	Quantity for Each Class of 32	Reusable in Each Class	Used in Activity
Adding machine tape, 5 meters	Long paper strips		1	No	7
Ball of string	Thread, yarn		50 meters	No	7, 10
Meter sticks	Yardsticks	1		Yes	7, 8
Envelopes	Folded paper	1		No	8
Construction paper	8 in. X 12 in. paper	2		No	8
Common objects for sorting	Paper clips, buttons, erasers		20	Yes	9
Dice	Number cubes	1 or 2		Yes	10

Laboratory Equipment

Table A.3—Laboratory Equipment.

Material	Substitutions or Alternatives. Optional Items Are Indicated.	Quantity per Pair, Team, or Center	Quantity for Each Class of 32	Reusable in Each Class	Used in Activity
Microscopes	Show video		4 to 16	Yes	2, 3
Glass slides	Plastic slides		4 to 24	Yes	2, 3
Cover slips			4 to 24	Yes	2, 3
Newspaper	Old handouts		1 page	No	2
Thread: red, yellow, blue	Any three colors		Spool	Yes	2
Feather			1	No	2
Rug fibers	Coarse material		A few	No	2
Trays			1 to 4	Yes	2
Hand lenses			4 to 16	Yes	3
Medicine droppers			5 to 6	Yes	3
Culture dishes	Shallow containers		1 to 4	Yes	3
Buckets	Big containers		4 to 8	Yes	3, 4
Dish soap			1	Yes	3, 4
Towels	Paper towels		1 to 4	Yes	3
Erlenmeyer flasks			10	No	4
Clamps	Clothes pins		10	No	4
Single-holed rubber stoppers			10	No	4
2 in. glass tubes			10	No	4
Grow lamp	Sunny window sill		1	Yes	4
Animal skeletons	Optional		Varies	Yes	8

Audiovisual Equipment

Table A.4—Audiovisual Equipment.

Material	Substitutions or Alternatives. Optional Items Are Indicated.	Quantity per Pair, Team, or Center	Quantity for Each Class of 32	Reusable in Each Class	Used in Activity
VCR, Monitor, Video	Black-line masters		1	Yes	3, 7
Overhead Projector	Xerox illustrations as handouts			Yes	2, 3, 5, 8
Overhead markers	Grease pencils		1	Yes	2, 3, 8

Ordering Information

"Living" materials may be ordered from Carolina Biological Supply Company. There are several alternative suppliers of biological materials.

Western United States, Alaska, and Hawaii order from
Powell Laboratories:
Call Toll-Free 800-547-1733
FAX 503-656-4208

Eastern United States order from
Carolina Biological Supply Company
Call Toll-Free 800-334-5551 (North Carolina customers call 800-632-1231)
FAX 919-584-3399

Teacher Background Information

You do not need to have all the answers to begin teaching your students about SETI! However, many teachers have asked for more information about the various topics presented in this unit. Therefore, the following notes about each of the missions are included. Please keep in mind that this book is written at an adult level, and it is not intended to be *read* to students. Enjoy!

Mission 1: Your SETI Academy Medical File

This mission was designed to offer students the opportunity to join the SETI Academy in an official manner while they begin to think about characteristics and traits of the human species as well as those of other possible intelligent species. The SETI Academy is totally fictional. However, students might like to know that all of the scientists who are pictured in the "Mission Briefings" logbook sheets are real people—real working scientists and teachers who have agreed to serve on the honorary staff of our imaginary academy.

Though the SETI Academy is a fictional device, the SETI *Institute* is a real establishment, as described in the text of this guide. Jill Tarter and Tom Pierson are real people who work for the SETI Institute.

Mission 2: Using a Microscope

This mission is a pre-laboratory activity that teaches the basics of microscope use. The activity is self-explanatory. No specifications are given for the microscopes. This is because the availability and cost of microscopes varies so greatly. If you purchase or borrow microscopes, be sure to choose sturdy "student-quality" equipment. As microscopes tend to be delicate, durability is more important than high degrees of resolution and magnification. The lesson asks for three magnifications. Because of microscope cost, some teachers have successfully used a single-objective microscope, a hand lens, and the naked eye. This works, of course, but higher-powered microscopes add to the experience.

Mission 3: Ancient Life-Forms

Culturing specimens is an exciting and rewarding experience that enriches student understanding of this mission. Note that the same cultures will be used again in mission 4, so plan to teach mission 4 as soon as possible after mission 3. Also, be sure to keep the specimens alive from one mission to the next.

Here we describe a few fairly simple ways to cultivate your own specimens, with or without class involvement. Carolina Biological Supply Company offers many suggestions and hints about growing your own cultures in a series of pamphlets. When teachers know the date they will be presenting a certain mission, they should call the supply company immediately to determine dates of delivery. The following information about how to grow and care for specimen cultures was compiled from these pamphlets, as well as from the feedback of our field test teachers.

Specimens

Cyanobacteria (blue-green bacteria)

Cyanobacteria belong to the phylum Eubacteria, which includes *Spirulina major*. They are single-celled or colonial; each cell is a simple cell without a nucleus. It is recommended that cyanobacteria be purchased from a biological supply company. Do not omit the cyanobacteria; they are critical to the next mission, and they represent an important primitive form of life. The water or walls of a well-established aquarium are likely to include cyanobacteria mixed with green algae. Because both are photosynthetic, they will thrive in well-lit conditions. However, cyanobacteria have a tendency to bleach when kept under high-intensity lights. All varieties should be kept in indirect light. For long-term preservation, cyanobacteria is best grown on a soil-water medium (see "Soil-Water Medium" below).

Green Algae

Green algae belong to the phylum Chlorophyta, which includes spirogyra, sea lettuce, and other plants having a pronounced green color. Green algae are true plants with complex cells that include a nucleus. These algae are difficult to grow in any media other than a soil-water medium (minus the calcium carbonate). To collect your own specimen, follow the directions in mission 3, or select a small sample of dried soil from the waterline of a pond, stream, lake, and so on and place in a small dish of sterile water under a bright light. Many different algae and protists will emerge within 24 hours.

Protists

Protists are microscopic, single-celled organisms; each cell is a complex cell with a nucleus. They belong to the phylum Protista, which includes the most primitive forms of life. They are neither plants nor animals: They are protists! They can be classified according to their means of locomotion. Amoebas are a type of protist that move with pseudopodia (false feet) that they put forth from any part of their "body" to engulf food. Flagellates, such as Euglena and Volvox, move about by means of a whip-like appendage, and ciliates, such as Paramecium, Stentor, and Vorticella, have short, eyelash-type hairs (cilia) bordering the edge of the cell body that beat in unison to move the organism through water.

Planaria

Planarians are invertebrate organisms that belong to the phylum Platyhelminthes (flatworms). They are true multicellular animals. Animal behaviorists consider flatworms the simplest animals that display an ability to respond to and learn from simple conditioning stimuli. They can master a two-choice maze. Freshwater planarians live on the underside of rocks in streams, creeks, and ponds. They avoid strong light and are more active at night than during the day. If you live near a clean body of water, you can collect planarians. Take a small piece of liver to the stream. Put it under a rock. Wait a few hours, or even days. If planarians are present, they will come.

If you purchase planarians, they need not be fed for about a week. When transferring them from their shipping container, use a pipette after you dislodge them from the sides or bottom. It is best to keep them in shallow trays of fresh spring water (not the bottled spring water from the supermarket). Use only spring or unpolluted pond water. The spring water they are kept in gets dirty quickly and should therefore be changed every day, if possible. Do not use soap or detergent to clean the bowl or tray in which the planarians are kept. Just use your finger to wipe away any slime from the sides or bottom before you pour off the dirty water.

Be cautioned that extreme temperatures of hot or cold can harm planarians. The cool side of room temperature, 21-23 degrees Celsius, is the acceptable temperature range in which to keep brown planarians (black planarians require temperatures slightly cooler, 16-18 degrees Celsius). Feeding planarians is described in mission 3.

Planarians are often used in regeneration experiments. If one is cut into pieces, each piece may regenerate, if it is large enough. Students can cut off the "head" and watch a new head grow back.

Bacteria

If our eyes were able to see as well as a microscope, we would notice bacteria almost everywhere. They grow in water, air, soil, foods, and in the tissues of plants and animals. Bacteria are single-celled, prokaryotic (a cell that lacks a membrane-bound nucleus) organisms. Most microorganisms do little harm to humans. In fact, many are essential for our health and even for our continued existence on Earth. We have symbiotic bacteria living in our intestines, producing vitamins for us. Prokaryotic organisms grow rapidly and are relatively simple structurally. Bacteria come in three basic cell shapes: sphere-shaped, or coccus; rod-shaped, or bacillus; and spiral-shaped (comma-shaped), or spirillum. If you grow your own bacteria as described in mission 3, use tongs to place the bacteria-laden potato into a shallow dish. Strongly caution students to use care when preparing a slide of bacteria for observation.

Soil-Water Medium

This can easily be made with simple materials. Sprinkle a pinch of calcium carbonate (the primary ingredient in some stomach antacids and chalk) over the bottom of a Petri dish, or a small, sterilized, plastic container, and cover with about one-half inch of sterilized, commercial potting soil. Next, fill the Petri dish or sterile container to three-quarters full with distilled spring water or pond water, and then cap the dish. Steam the dish for two hours on two consecutive days. Note: When making a soil-water medium for green algae, omit the calcium carbonate.

Standard Methodology for Culturing Bacteria

It is important to begin with sterilized Petri dishes for growing microbes. There are microbes in the air, in water, on our hands, on gloves, in paper towels—nearly everywhere. These microbes can contaminate exposed Petri dishes. We recommend purchasing commercial, easy-to-use, pre-sterilized, disposable Petri dishes. We also provide instructions for sterilizing your own Petri dishes. This should be adequate for class use, although teachers report problems with contamination and setup time for the media.

Purchase sterile dishes. Sterile, disposable, polystyrene Petri dishes sealed in plastic sleeves may be ordered from Carolina Biological Supply Company. They remain sterile if stored in the unopened sleeves.

Sterilize your own dishes. Use an autoclave if you have access to one. Hospitals and colleges often have autoclaves. If not, use one of the following methods:

- Microwave: Using dry, clean plastic or glass Petri dishes, close the lids and place in the microwave. Microwave on high for three minutes. These dishes will remain uncontaminated, at least overnight, if left unopened.
- Oven: Heat the Petri dishes with their covers on in an oven at 150 degrees for one hour. Turn off the oven and let the dishes cool, leaving the covers on.
- Boiling water: Pour boiling water into the Petri dishes, let it stand for a minute, pour out the water, rinse the insides of the covers with boiling water, and then put the covers back on.

Media for Feeding Bacteria

Whenever you grow microbes in Petri dishes, you must provide an adequate food source. We recommend purchasing a commercial, easy-to-use product, such as Sterigel. We also provide instructions for a homemade nutrient agar that will be adequate for class use. The homemade nutrient agar often does not set well, and it becomes contaminated easily. Follow directions closely. A trial run is advised.

Purchase a prepared medium. Sterigel Instant Medium may be ordered from Carolina Biological Supply. This medium is sterile, and it usually sets up quickly enough to be prepared in the classroom as it is needed. Follow the enclosed instructions. Sterigel Instant Medium comes in two parts: a liquid and a powder. Once you have mixed the two parts together, they will gel in about 20 minutes. This gel cannot be melted later. You must make up no more than enough for one class at a time.

Make your own medium. Recipe for a nutrient medium for 30 dishes:

- 3 quarter-ounce packets of Knox unflavored gelatin
- one box Sure-Jell fruit pectin
- access to a refrigerator to set the gel
- Pyrex measuring cup, or cup to hold boiling water
- roll of plastic wrap
- one stainless steel soup pan

Knox unflavored gelatin and Sure-Jell fruit pectin are available in supermarkets. This medium must be prepared the day before class, and it must be refrigerated at school to become solid enough for students to use or transport. Use a microwave or an oven. To prepare the nutrient gelatin, bring about six cups of water to a boil. Sterilize the Pyrex cup and the soup pan. Put three quarter-ounce packages of Knox gelatin into the pan. Sprinkle three-quarters of a teaspoon of Sure-Jell over the Knox gelatin. Use the Pyrex cup to add three cups of boiling water. Mix by swirling for a few seconds. Pull a fresh sheet of plastic wrap over the top, to seal the pan. Let this pan stand at room temperature until the next day's class. The gelatin will be a thick, syrupy liquid. Put the Petri dishes into the refrigerator (not the freezer section). The nutrient gelatin should be permanently solid after about two hours.

In addition to growing your own cultures, you might need to order one or two specimens from a biological supply company. These have the major advantage of being clearly identified. Short-term preservation of these cultures is recommended. If you order cultures, be sure to give a specific use date. Cultures can be kept in good condition for a few days by loosening the caps of the tubes and placing them in a cool, dimly lit area (not a refrigerator). If you need to keep specimens for more than four days, transfer them into sterile spring water (water from a spring or pond, *not* bottled spring water from the store) where they can be kept in usable condition for up to two weeks.

Mission 4: Who Changed Earth's Atmosphere?

The theory about the origins of life most pertinent to the topics presented in the SETI guides is stated best by the following summary about the chemical evolution of life.

Life developed gradually through a series of continuous changes in the chemical systems that existed on primitive Earth. Primitive Earth was different from Earth today. There were more storms and volcanic activity. Most importantly, the early atmosphere was very low in oxygen (some scientists say *totally lacking* in oxygen). We, or any other organism that relies on oxygen for aerobic respiration, could not have survived there. However, the abundant chemicals and available energy almost certainly caused the formation of many organic chemicals; the oceans became a sort of "primordial soup," which included amino acids and simple sugars.

The first living cells possibly arose in this fertile, tepid, shallow sea of organic soup on early Earth. Simple chemicals such as monomers could have combined into more complex polymers. It is hypothesized that the first prebiont or protobiont (a pre-cell) developed a membrane that separated its internal environment from its external environment, structural organization, and the ability to maintain the membrane or cell wall that separated its insides from the rest of the organic soup. It also became competent in converting chemical compounds available in the organic soup for food. At some point, it became able to replicate itself or reproduce, with either RNA or DNA, and life was created.

It was inevitable that these heterotrophic forms would eat up all their food eventually, as they reproduced but did not make new foods. As the organic matter in the organic soup became scarce, some early cells began to use radiant energy from the Sun, combined with raw materials of carbon dioxide and water present in the organic soup, to produce fuel for themselves in a process known as photosynthesis. One of the by-products of photosynthesis is oxygen. As photosynthetic cells produced more and more oxygen, the atmosphere started to change. Eventually, after two billion years, Earth's atmosphere became oxygen rich and able to support aerobic life-forms.

A more recent alternative hypothesis about the origin of organic compounds and the first life has been proposed by oceanographers and geologists who believe that ancient underwater vents of superheated water and magma along the seams between oceanic plates could be the source of primitive organic compounds and energy. The thriving ecology of undersea vents today is a testimony to their ability to maintain life, independent of solar energy.

Mission 5: Fossils!

Fossils are the preserved remains or impressions (including footprints) of biological specimens that lived in the distant past. Primarily, they consist of the bone structure of animals and the cellulose structure of plants, because organs and other soft tissues decompose not long after an organism dies (although they can form a mold in which rocks or minerals are deposited). There are several types of fossils. In one type, the hard parts of the organisms themselves remain. This accounts for shark teeth and dinosaur skulls and skeletons. In these cases, all the soft, organic material has rotted away. In a second type, the impressions of soft-bodied animals are left after the animals are covered by soft material. The Burgess Shale fossils are impressions of fragile, soft-bodied creatures like today's jellyfish that were covered with fine sediments. The actual bodies are long gone, but the impressions are quite detailed. In some cases, butterflies have been covered with ash from a volcanic eruption, leaving an exact image of the butterfly. Another type of fossil consists of insects and small arthropods that became embedded in tree sap, which itself fossilized, turning into amber. In the La Brea Tar Pits, in what is now west Los Angeles, tar bubbled up to the surface, forming pools. Many animals became trapped in the tar. In another type of fossilization, rock-forming chemicals replace the original organic materials. The Petrified Forest in Arizona has huge pieces of petrified wood from ancient trees that once grew there. Some types of preservation do not produce "true fossils": Woolly mammoths have been found freeze-dried, with the meat still edible, and human bodies have been recovered from peat bogs with traces of clothing still intact.

Students should be aware of the difference between sedimentary rock, which is a good source of fossils, and igneous and metamorphic rock, which is not. However, it is the igneous rock that is good for "absolute" dating, because it is formed at a point in time with a specific ratio of isotopes. This ratio then changes at a predictable rate. The dating of a fossil is accomplished by using both relative and absolute dating. For instance, if a fossil is found in a layer of sedimentary rock that is sandwiched between two layers of igneous rock, both of the igneous layers can be dated. This will give a range of ages: The fossil is younger than the lower rock, but older than the upper one.

Mission 6: Natural Selection

This mission is self-explanatory. If students have sufficient math background, you may wish to introduce a numerical analysis in class of the change brought about by natural selection. If students have any background in genetics, this is an ideal time to introduce the concepts of populations, gene pools, and the change in gene frequencies. The vast and complex topic of genetics is not introduced in this guide. The scientists at SETI, like Darwin, simply assume that there is a mechanism that causes children to resemble their parents.

You may wish to introduce the idea of *artificial* selection. Students should see the parallels between natural and artificial selection, as some "better" variations (cats with curly ears) get to survive and have more offspring (kittens with curly ears!), while "bad" variations (regular house-cat ears) may not get to survive or produce offspring. Over time, we get different breeds of cats, dogs, pigeons, and so forth. Students can easily see that dogs vary in appearance, but they have trouble extrapolating large changes from small ones. They could be reminded that the dog itself began as a wolf. Domestication of some wolves, long ago in human history, literally "created" the species called *dog*. The same can be said of cattle and other domestic animals. Point out that the critical difference is that, in artificial selection, there is a human intelligence making conscious choices. Of course, this is not true of natural selection—as far as science can demonstrate.

Students wonder, Is natural selection still occurring today? Indeed it is. Scientists are investigating such genetic changes all over the world, from fruit fly speciation in Hawaii to tree snails in the Everglades of Florida. Also, many laboratory experiments have demonstrated natural selection and the adaptation to various temperatures, chemicals, and so forth.

Mission 7: A Timeline for the Evolution of Life

The descriptions beside each picture in the videotaped image show are a good summary of the major events in the evolution of life. There are also many "minor" events that are well known to paleontologists not presented here. We provide only a cursory summary of a vast and complex topic.

The first cells were prokaryotic, simple cells without a nucleus. They evolved approximately three billion years ago. Eukaryotic cells with a defined nucleus containing genetic material evolved next, approximately 1.5 to 2 billion years ago. Once these cells developed, biological evolution was faster. Eukaryotes developed true sexual reproduction, which sped things up a good deal via genetic recombination. The first multicellular organisms evolved from unicellular eukaryotes about one billion years ago. These early life-forms gave rise to all the plants and animals that exist on Earth today; all the "higher" life-forms have eukaryotic cells. Table A.5 presents a brief summary of the major events in the later evolution of life on Earth.

Mission 8: Tracing Family Trees

The information required to understand this activity is a further explanation of the fossil record, and students will use parts of this as they piece together a vertebrate family tree. Through the study of fossils, scientists are able to learn about the structures of ancient life-forms and compare them to the structures of modern-day organisms. The fossil records also allow scientists to theorize about the environments in which organisms lived as well as build a time frame for the evolution of life-forms that shows the connectedness of a variety of species. Evolution of living organisms is a basic biological theory that explains how they change through time. Entire populations, not individuals, evolve as natural selection kills the unfit and allows the fit to survive.

The first vertebrate animals to evolve on Earth were jawless fish; this occurred during the Ordovician period, between 430 and 500 million years ago. The Silurian period (395-430 mya) that followed included the evolution of fish with jaws. Next in geologic time came the Devonian period (345-395 mya), which saw the move of vertebrates onto land. This was when a lobed-fin fish abandoned its life in shrinking supplies of water to search for other water sources. This very small, yet major step in evolution, ultimately gave rise to amphibians, vertebrates that live part of their life in the water and part on land, and to reptiles, another cold-blooded vertebrate. Reptiles can be found in the fossil record beginning in the Carboniferous period (280-345 mya). During the Permian period (225-280 mya), reptiles began to assume the role of dominant life-form on Earth, and by the following period, the Triassic (190-225 mya), despite several mass extinctions involving many groups, the first dinosaurs had become dominant and the first tiny mammals had increased in number. During the Jurassic period (136-190 mya), dinosaurs had diversified greatly; they remained the dominant vertebrates along with the first birds and archaic mammals. The Cretaceous period (65-136 mya) was characterized by continued diversification of dinosaurs, continued evolutionary changes

Table A.5—Timeline for the Later Evolution of Life.

Geologic Time Period	Million Years Ago	Notable Events
Odovician	430 - 500	Diversification of invertebrates; origin of jawless fish; mass extinction at end of period
Silurian	395 - 430	Diversification of jawless fish; land plants and arthropods appear
Devonian	345 - 395	Origin and radiation of bony and cartilaginous fish; origin of amphibians, chambered molluscs and insects; mass extinction at end of period
Carboniferous	280 - 345	Extended forests, amphibians radiate; first reptiles
Permian	225 - 280	Reptiles diversify; amphibians decrease; continents form Pangaea; glaciations; mass extinction at end of period
Triassic	190 - 225	First dinosaurs and mammals; coniferous trees dominate; continents begin to drift; mass extinction of marine organisms at end of period
Jurassic	136 - 190	Dinosaur varieties increase; birds and ancient mammals first appear; land masses continue to drift
Cretaceous	65 - 136	Continents widely separated; dinosuars diversify; flowering plants (angiosperms) and mammal species begin to radiate; mass extinction at end of period
Tertiary	2 - 65	Mammals, birds, angiosperms continue to diversify; continents almost reach current positions; drought conditions in mid-period
Quaternary	Present - 2	Multiple glaciations; extinctions of large mammals; evolution of *Homo sapiens*; rise of civilization

among mammals, and another massive extinction of many life-forms, both plant and animal. This extinction has been credited to climactic changes that occurred after a very large meteor collided with Earth approximately 65 million years ago. Next, the Tertiary period occurred (2-65 mya), at which time vertebrate mammals and birds continued to evolve. Finally, the Quaternary period began two million years ago, during which many large mammals became extinct, repeated glaciations occurred, and the evolution of the planet's dominant life-form, the hominid family, also occurred.

Mission 9: What Organism Do You See?

The guessing game that comprises this mission is designed to get students thinking about the major characteristics of vertebrates, and to get students to learn that all living organisms can be classified. The "Organism Classification Key" in the student logbooks was structured according to biologic guidelines that relate physical characteristics of living things. As most familiar, observable species belong to either the plant or the animal kingdom, if students follow the questioning procedure on the classification key they should have an enjoyable, educational experience with this activity. The major divisions are:

Kingdom. All Earth organisms are organized into five major groups called kingdoms. One kingdom is Animalia (the animal kingdom), which is one of the kingdoms students examined in the game. The other kingdoms are the plants; the bacteria, which includes the cyanobacteria (the "O2 Culprit"); the protists (seen in the microscope lab with the pond water); and the fungi, a kingdom that includes mushrooms and yeast.

Teacher's Note: The term invertebrate *is mostly used in a much more general sense to describe any animal without a backbone. This grouping includes all members of every animal phylum except Chordata, plus a few invertebrate Chordates, like the lancelet and the tunicate.*

Phylum. Each kingdom is subdivided into smaller groups, each of which is called a phylum. The plural is "phyla." Sea stars belong in the phylum Echinodermata. Human beings belong in the phylum Chordata.

Subphylum. Each phylum is subdivided into smaller groups, each of which is called a subphylum. The plural is "subphyla." The phylum Chordata is subdivided into animals with backbones, which are in a subphylum Vertebrata (vertebrates), and animals without backbones, which are in the subphylum Invertebrata (invertebrates).

Class. Each subphylum (or phylum) is subdivided into smaller groups, each of which is called a class. For example, Vertebrata are broken down into several classes including Chondrichthyes (cartilaginous fish), Teleostomi (bony fish), Amphibia (amphibians), Reptilia (reptiles), Aves (birds), and Mammalia (mammals). Humans are mammals.

Order. Each class is subdivided into smaller groups, each of which is called an order. Orders have many important characteristics in common. For example, humans, monkeys, chimpanzees, gorillas, and orangutans are all in the order Primates, which is one order of the class Mammalia (mammals).

Family. Each order is subdivided into smaller groups, each of which is called a family. One family of Primates is called Hominidae. The hominid family includes several human-like species, all of which are extinct except for our own.

Genus and Species. Each organism has a genus (generic) name and a species (specific) name. For example, humans are *Homo sapiens*. Humans and gorillas are different species of Hominidae. Plants and animals that can breed with each other are members of the same species.

Teacher's Note: This is a simplified scheme. It leaves out subclasses, suborders, subfamilies, and subspecies.

There are many types of vertebrate animal life and many types of invertebrate life. Biologic guidelines that relate physical characteristics to living things are as follows:

Vertebrates—Animals with backbones.

Fish—Any of a number of cold-blooded, aquatic vertebrates having fins, gills, and a streamlined body. Today, this includes bony fish, such as trout or guppies; cartilaginous fish, such as sharks; and jawless fish, such as lampreys.

Amphibians—Any of a number of cold-blooded, smooth-skinned vertebrate organisms such as frogs, toads, and salamanders. They characteristically hatch from soft, jelly-like eggs, spend the first phase of their lives as aquatic larvae that breathe via gills, and metamorphose into an adult with air-breathing lungs. Note that this means that these creatures have the genes for gill *and* lung use.

Reptiles—Any of a number of cold-blooded, usually egg-laying vertebrates such as snakes, lizards, crocodiles, turtles, and dinosaurs. They usually have an external waterproof covering of scales or horny plates and breathe through lungs. Their eggs have a waterproof leathery shell.

Birds—Any of a number of warm-blooded, egg-laying vertebrates with forelimbs modified to form wings. These eggs have shells.

Mammals—Any of a number of warm-blooded vertebrates, consisting of more than 4,000 species, including humans. Mammals are distinguished by self-regulating body temperature, hair or fur, and milk-producing glands in the females. Only the most primitive (monotremes) still lay reptile-style eggs. Marsupials bear tiny young and carry them in a pouch. Placentals give birth to well-formed, live young.

Invertebrates—Animals without backbones include many groups such as Porifera (sponges), Platyhelminthes (flatworms), Annelida (segmented worms), Arthropoda (crabs, insects, and spiders), Echinodermata (starfish, sea urchins, and sand dollars), and Chordata (some invertebrates such as sea squirts; all vertebrates).

Mission 10: Inventing Life-Forms

There are certain environmental challenges that all life-forms face and must solve. The variety of the solutions seen in animals on Earth is stunning. This is a good mission to relate to issues in ecology and animal behavior.

Mission 11: Creating Your Extraterrestrial's Family Tree

This mission tests whether students can apply what they learned to create a plausible simulation of evolution. It ties together many of the ideas from previous missions. The challenge here is to allow creativity, but to keep it within reasonable bounds.

Mission 12: Mission Completed!

This mission is actually more like a final exam. No new material is presented, and the only activity is answering questions. It is intended for use as an assessment of how much students have learned.

Resources

Resources for SETI Academy Cadets

Popular interest and continuing discoveries about dinosaurs and other extinct life-forms have encouraged production of resources for children that focus on evolution. Many have excellent pictures, texts, and even interactive components that can help stimulate interest, provide visual models, and encourage further research as students proceed through their investigations as SETI Academy cadets.

The following is a list of resources that will be helpful in book two of the *SETI Academy Planet Project*. It is organized by topic to help identify those materials that will be the most helpful at any given stage of the SETI investigation. No doubt, you will find other resources in local libraries, stores, and media centers as you explore and as more resources become available.

SETI

Darling, David. *Other Worlds: Is There Life Out There?* Minneapolis, Minn.: Dillon Press, 1985.

One of the Discovering Our Universe series, this book explores the question that SETI cadets will investigate: Is there life on other planets? The author answers common questions and describes the search for life on other planets in our solar system. He concludes by giving evidence that suggests the possibility of life in other stellar systems and the means by which we might detect it.

Evolution of the Solar System, Earth, and Life on Earth

Note: Many of the following books were used in the first book of the *SETI Academy Planet Project* (*The Evolution of a Planetary System*). They will be useful again in this book.

Benton, Michael. *The Story of Life on Earth*. New York: Warwick Press, 1986.

Michael Benton's book traces the origins and development of life on Earth. It focuses on the use of evidence from ancient rocks to determine the age of Earth, how continents are moving, and what life-forms have existed. It also includes useful diagrams of the arrangement of the continents during key eras.

Branley, Franklin. *The Beginning of the Earth*. New York: Crowell, 1972.

A simple version of the formation and evolution of Earth. Use this book for its lively illustrations or read it without showing illustrations so that students can make their own storyboard drawings of Earth's early formation.

Burton, Virginia. *Life Story*. Boston: Houghton Mifflin, 1962.

Although this classic is often used with young children, its five-act-play format with prologue and epilogue introduces life in the Milky Way Galaxy in an easily told story. It can be used to support information in the SETI video image show, as a springboard for discussion of questions about the evolution of Earth and its life, as a format for a student play, or just as enjoyable literature.

Gonick, Larry. *The Cartoon History of the Universe*. New York: Doubleday, 1990.
A tongue-in-cheek cartoon history that appeals to young adolescents and adults, this paperback volume is filled with facts. This is the kind of book that gets lost in a desk because the borrower won't give it up.

Minelli, Giuseppe. *The Evolution of Life* . New York: Facts on File, 1986.
Another picturesque history of the formation of our planetary system and the development of life on Earth.

Rand McNally, ed. *Children's Atlas of Earth Through Time*. Chicago: Rand McNally, 1990.
This is a fine resource book for upper elementary students. It clearly describes the formation of the universe starting with the "Big Bang," the development of the Milky Way Galaxy, and our solar system. It goes on to outline the evolution of Earth and its life through geologic eras and periods using specific life-forms as examples.

Saville, David. *The Evolution of the World*. New York: Hyperion Books, 1991.
In 12 pages of text and revolving pictures, children can experience the major events in Earth's evolutionary history.

Schwartz, David. *How Much Is a Million*? New York: Lothrop, Lee, and Shepard, 1985.
Imagining the immensely long times required for the evolution of life on Earth poses a real problem for students. This book, combined with a number of "millions collection" activities from mathematics sources, helps students visualize these large numbers.

Teacher Resources

Books and Articles

Berra, Tim. *Evolution and the Myth of Creationism*. Stanford, Calif.: Stanford University Press, 1990.
This is mostly a short course in evolutionary theory, with some references to Creationist tactics and arguments. A good refresher or introduction to evolution.

Gallant, Roy. *Before the Sun Dies: The Story of Evolution*. New York: Macmillan, 1989.
Although this is considered a children's book, it is a complete, well-written volume on the evolution of our solar system, Earth, and life that would be an excellent resource for a teacher about to plunge into this subject.

Godfrey, Laurie. *Scientists Confront Creationism*. New York: W. W. Norton, 1980.
Creationists use some odd arguments to make their case. These can seem very convincing to students, especially if they are used by parents and religious leaders. If you wish to be aware of the Creationist arguments in advance, this book shows each argument and how scientists refute it. If you have argumentative students, and if you wish to "confront" them on one point or another, consider reading this book.

Goldsmith, Donald, and Tobias Owen. *The Search for Life in the Universe*. Reading, Mass.: Addison-Wesley, 1980.
This is a complete reference explaining many of the scientific concepts that make a search for extraterrestrial life plausible. This is the adult version upon which the missions of the *SETI Academy Planet Project* is based.

McDonough, Thomas. *The Search for Extraterrestrial Intelligence: Listening for Life in the Cosmos*. New York: John Wiley, 1987.

Thomas McDonough has written a history of the search for extraterrestrial beings including the hoaxes, the connections with science fiction, the many discoveries, and the people who have made the search reality. The book is written in a humorous and accessible style.

Sobel, Dava. "Is Anybody Out There?" *Life*, September 1992, 60-64.

This *Life* magazine article begins with a quote by Jill Tarter's eight-year-old daughter that puts into perspective what her mother and other SETI scientists do every day and why they do it. Nice graphics and pictures of Jill Tarter and Frank Drake. Following the article is a one-page opinion by Arthur C. Clarke entitled "Why Is It Important?" that discusses the human need to know whether we are really alone.

Additional Activities

Stein, Sara. *The Evolution Book*. New York: Workman, 1986.

This book provides many observations, experiments, projects, and investigations for children ages 10-14.

Resource Centers

SETI teacher resource guides, posters, and videos will be available from the following source as they are produced. Presently, there are two posters produced by SETI artist Jon Lomberg. For information write to:

SETI Institute
Attn.: S. Shostak
2035 Landings Drive
Mountain View, CA 94043

For additional NASA teacher materials, including available videos and information about educational programs, contact the following, or the NASA Teacher Resource Center nearest you:

Ames Research Center
Mail Stop T025
Moffett Field, CA. 94035
415-604-3574

Films, Videos, Laser Discs, and Computer Software

Consult your local audiovisual sources and/or the NASA Teacher Resource Center in your area for additional audiovisual materials.

Sim Earth. Maxis Software.

This is a popular computer simulation that allows players to manipulate the conditions of Earth's environment. Students learn through experience the delicate balance that exists between each of the variables that allow Earth to remain a habitable planet. Call 800-23-MAXIS for more information.

Black-Line Masters

The following pages contain black-line masters that can be reproduced for overhead projection.

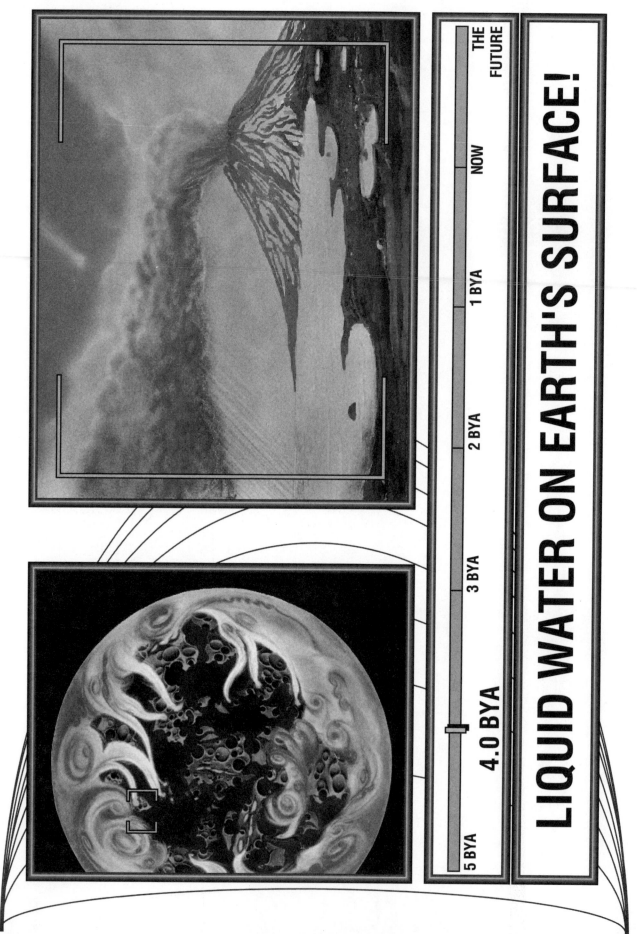

LIQUID WATER ON EARTH'S SURFACE!

5 BYA — 4.0 BYA — 3 BYA — 2 BYA — 1 BYA — NOW — THE FUTURE

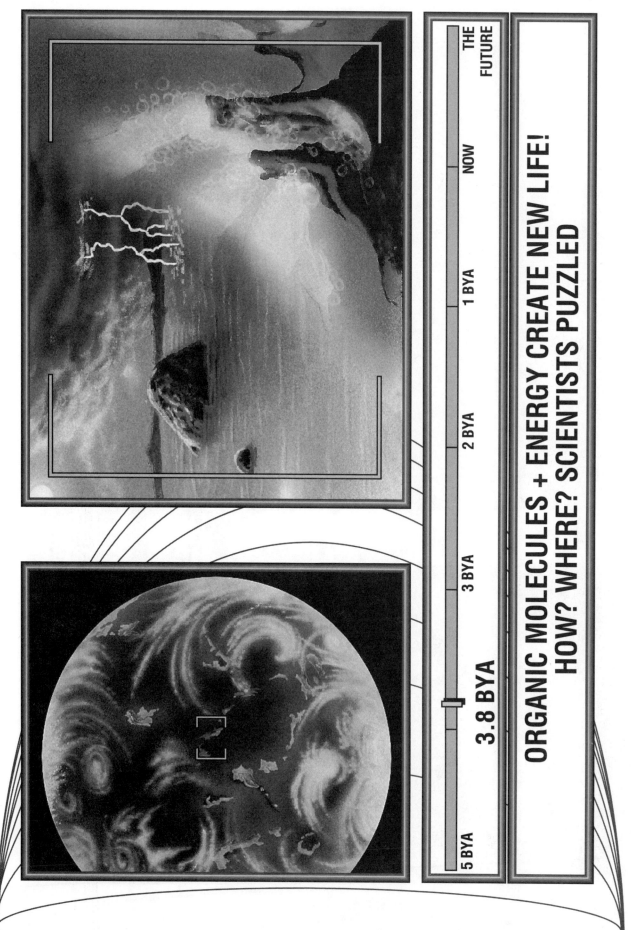

5 BYA — 3.8 BYA — 3 BYA — 2 BYA — 1 BYA — NOW — THE FUTURE

ORGANIC MOLECULES + ENERGY CREATE NEW LIFE! HOW? WHERE? SCIENTISTS PUZZLED

From How Might Life Evolve on Other Worlds? © 1995. Teacher Ideas Press. (800) 237-6124.

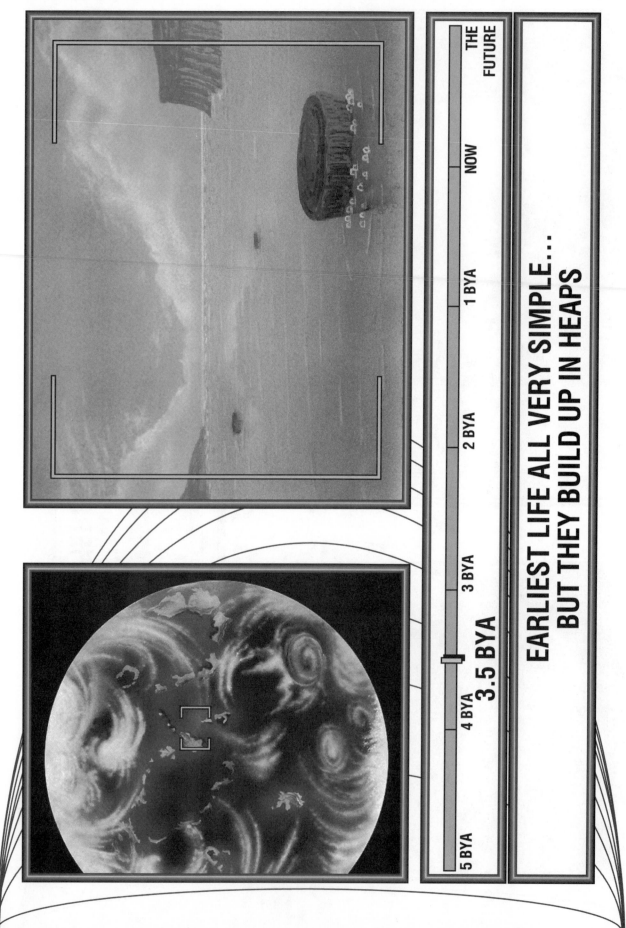

5 BYA 4 BYA **3.5 BYA** 3 BYA 2 BYA 1 BYA NOW THE FUTURE

EARLIEST LIFE ALL VERY SIMPLE... BUT THEY BUILD UP IN HEAPS

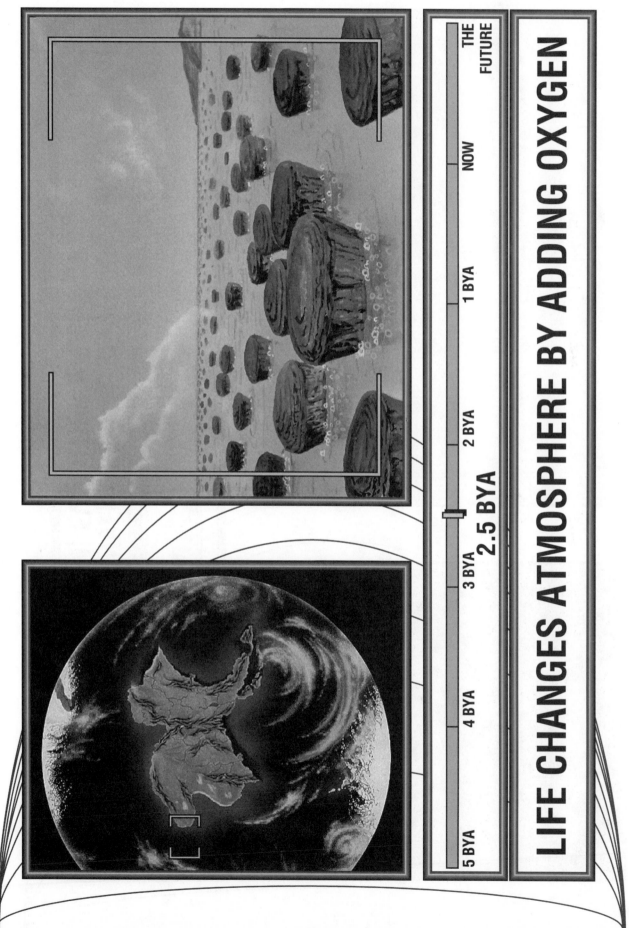

LIFE CHANGES ATMOSPHERE BY ADDING OXYGEN

5 BYA | 4 BYA | 3 BYA | 2.5 BYA | 2 BYA | 1 BYA | NOW | THE FUTURE

From How Might Life Evolve on Other Worlds? © 1995. Teacher Ideas Press. (800) 237-6124.

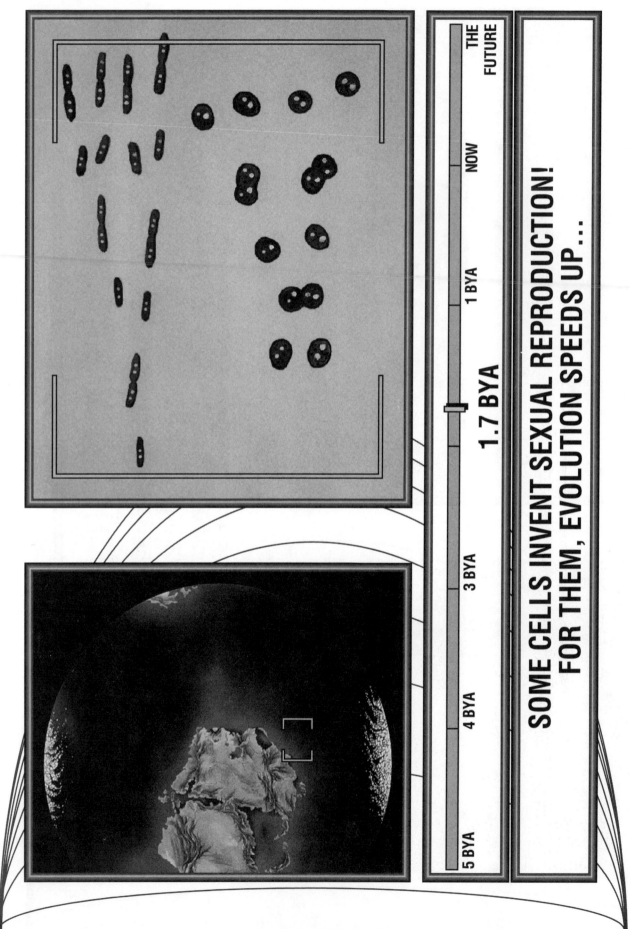

SOME CELLS INVENT SEXUAL REPRODUCTION! FOR THEM, EVOLUTION SPEEDS UP . . .

5 BYA   4 BYA   3 BYA   1.7 BYA   1 BYA   NOW   THE FUTURE

From *How Might Life Evolve on Other Worlds?* © 1995. Teacher Ideas Press. (800) 237-6124.

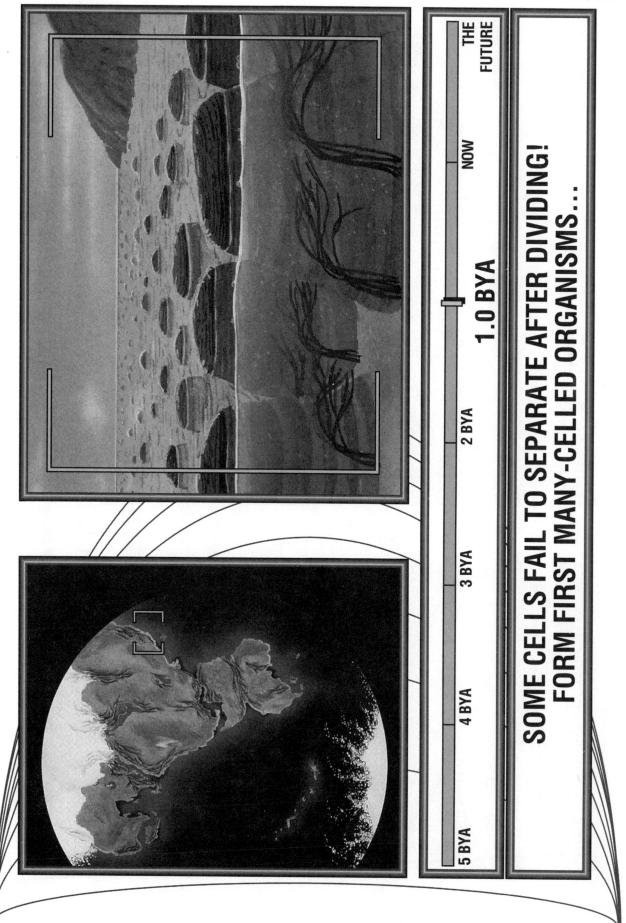

5 BYA · 4 BYA · 3 BYA · 2 BYA · 1.0 BYA · NOW · THE FUTURE

SOME CELLS FAIL TO SEPARATE AFTER DIVIDING! FORM FIRST MANY-CELLED ORGANISMS...

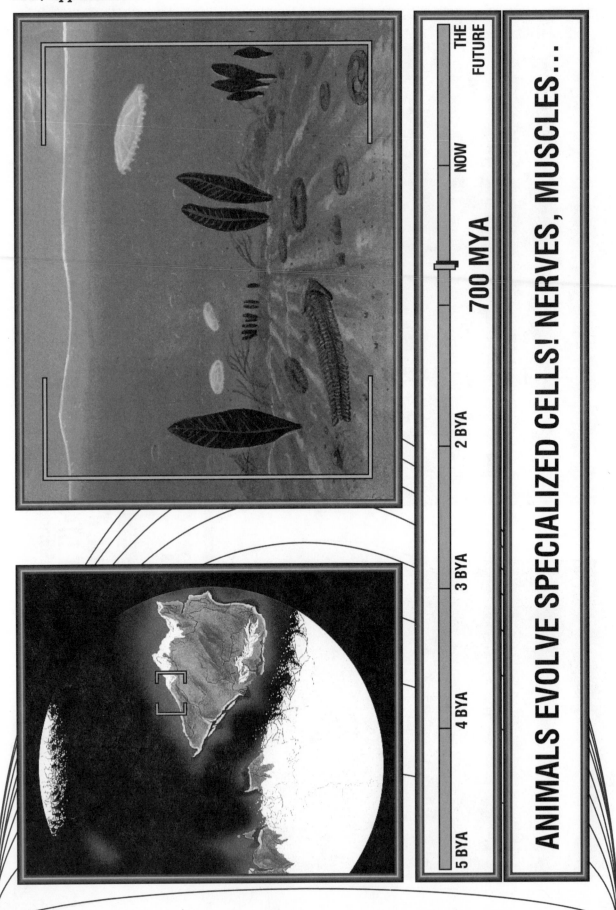

5 BYA   4 BYA   3 BYA   2 BYA   700 MYA   NOW   THE FUTURE

ANIMALS EVOLVE SPECIALIZED CELLS! NERVES, MUSCLES…

From *How Might Life Evolve on Other Worlds?* © 1995. Teacher Ideas Press. (800) 237-6124.

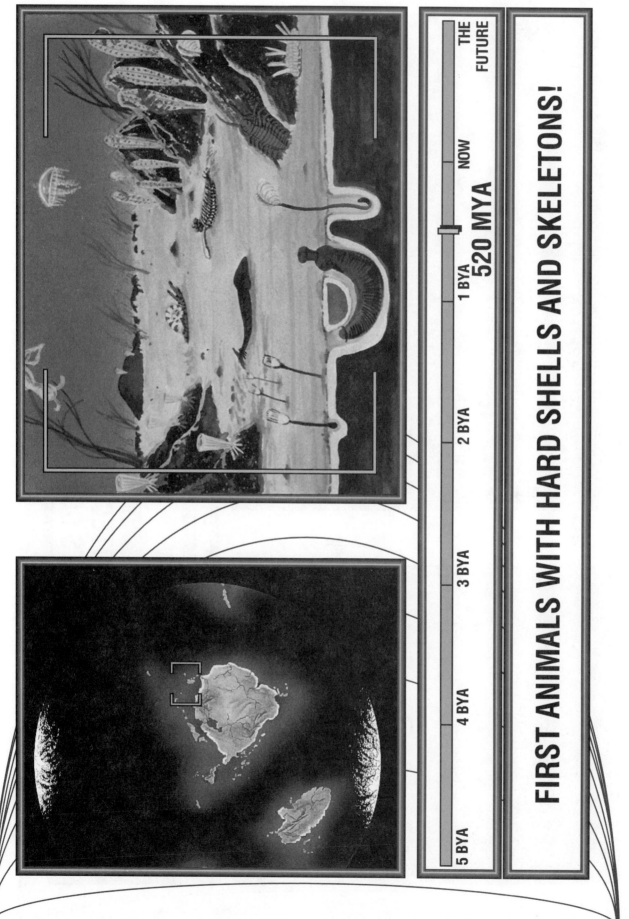

5 BYA   4 BYA   3 BYA   2 BYA   1 BYA   NOW   THE FUTURE

520 MYA

FIRST ANIMALS WITH HARD SHELLS AND SKELETONS!

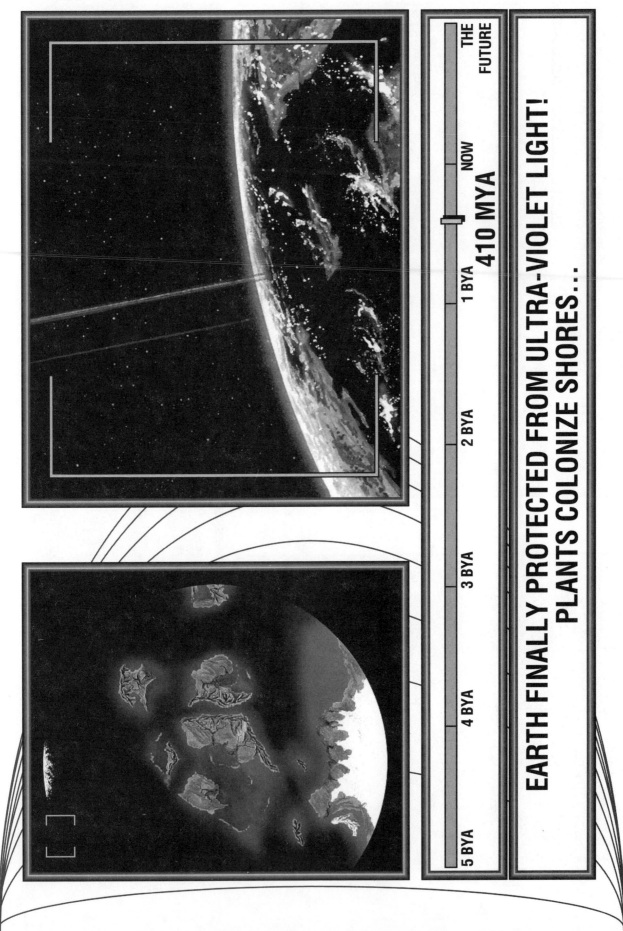

EARTH FINALLY PROTECTED FROM ULTRA-VIOLET LIGHT!
PLANTS COLONIZE SHORES...

410 MYA

5 BYA   4 BYA   3 BYA   2 BYA   1 BYA   NOW   THE FUTURE

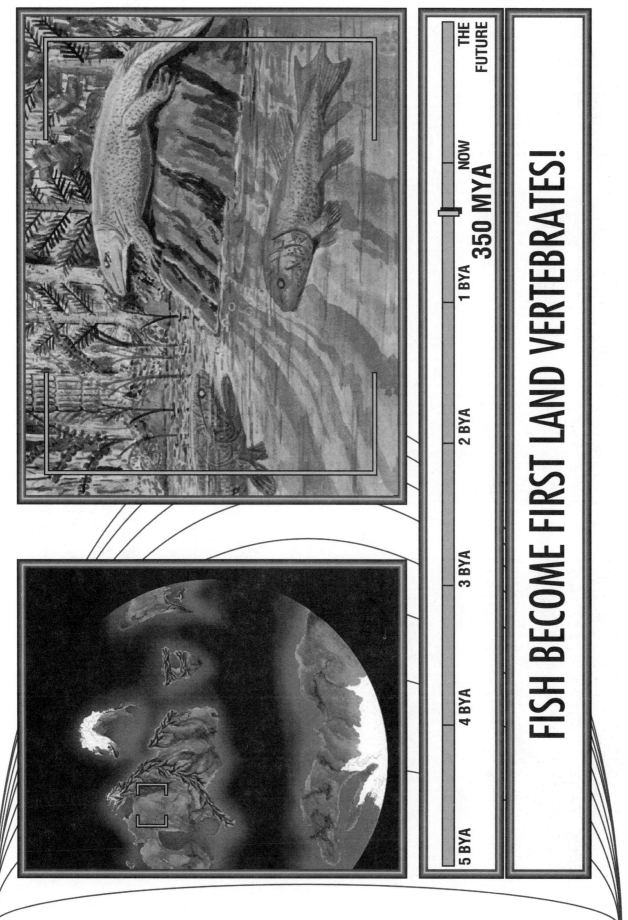

5 BYA 4 BYA 3 BYA 2 BYA 1 BYA NOW THE FUTURE

350 MYA

FISH BECOME FIRST LAND VERTEBRATES!

From How Might Life Evolve on Other Worlds? © 1995. Teacher Ideas Press. (800) 237-6124.

ABUNDANT LIFE ON LAND! AMPHIBIANS, INSECTS AND THEIR RELATIVES, PLANTS, FIRST REPTILES!!

300 MYA

5 BYA — 4 BYA — 3 BYA — 2 BYA — 1 BYA — NOW — THE FUTURE

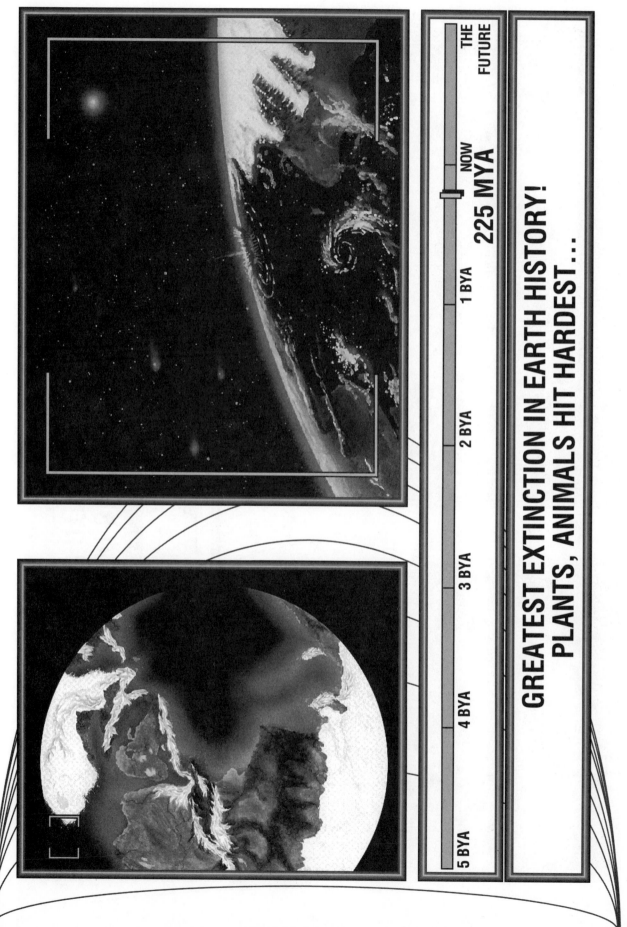

5 BYA   4 BYA   3 BYA   2 BYA   1 BYA   NOW   THE FUTURE

225 MYA

**GREATEST EXTINCTION IN EARTH HISTORY!
PLANTS, ANIMALS HIT HARDEST . . .**

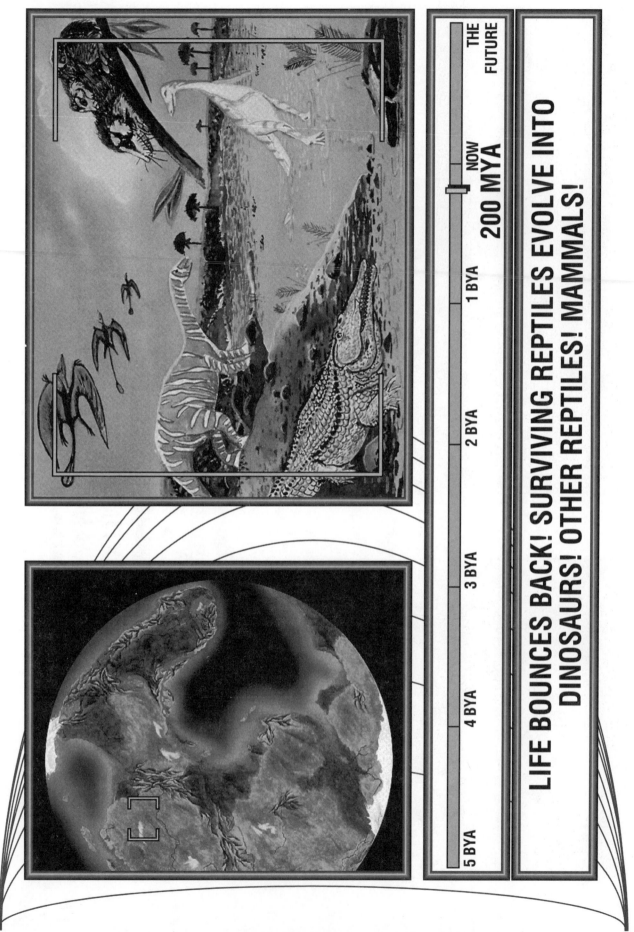

5 BYA   4 BYA   3 BYA   2 BYA   1 BYA   200 MYA   NOW   THE FUTURE

LIFE BOUNCES BACK! SURVIVING REPTILES EVOLVE INTO DINOSAURS! OTHER REPTILES! MAMMALS!

From *How Might Life Evolve on Other Worlds?* © 1995. Teacher Ideas Press. (800) 237-6124.

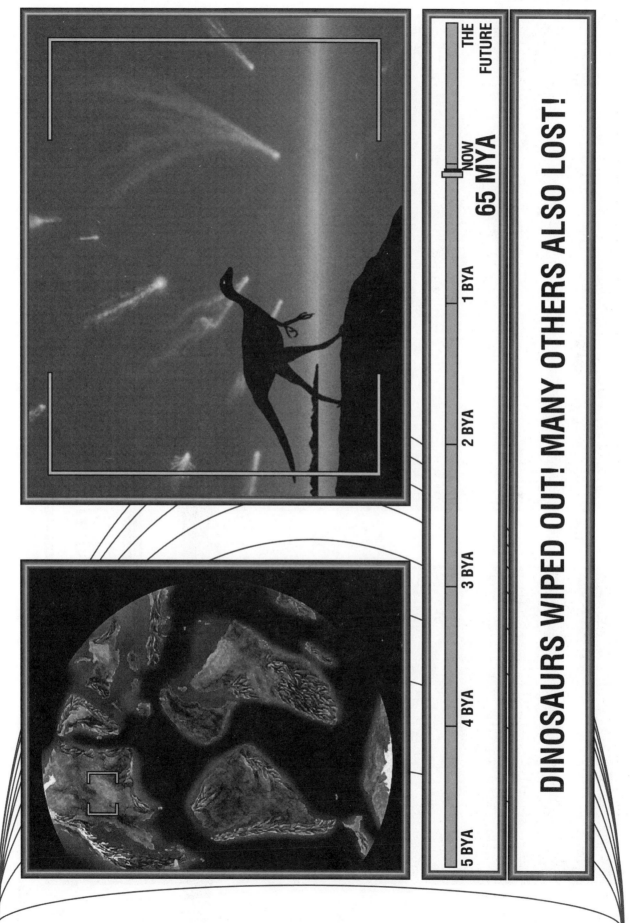

THE FUTURE

NOW
65 MYA

1 BYA

2 BYA

3 BYA

4 BYA

5 BYA

DINOSAURS WIPED OUT! MANY OTHERS ALSO LOST!

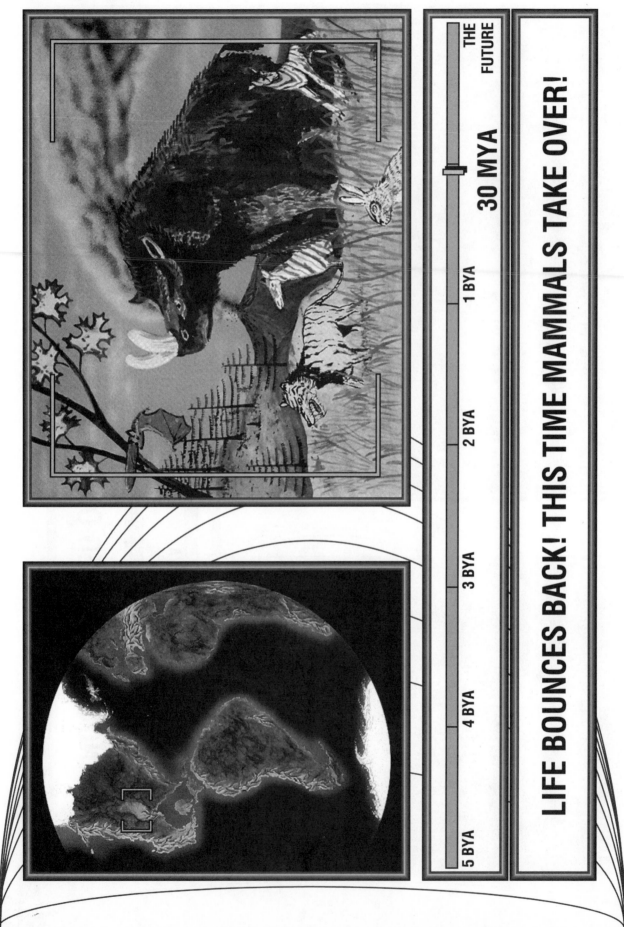

LIFE BOUNCES BACK! THIS TIME MAMMALS TAKE OVER!

5 BYA   4 BYA   3 BYA   2 BYA   1 BYA   30 MYA   THE FUTURE

From *How Might Life Evolve on Other Worlds?* © 1995. Teacher Ideas Press. (800) 237-6124.

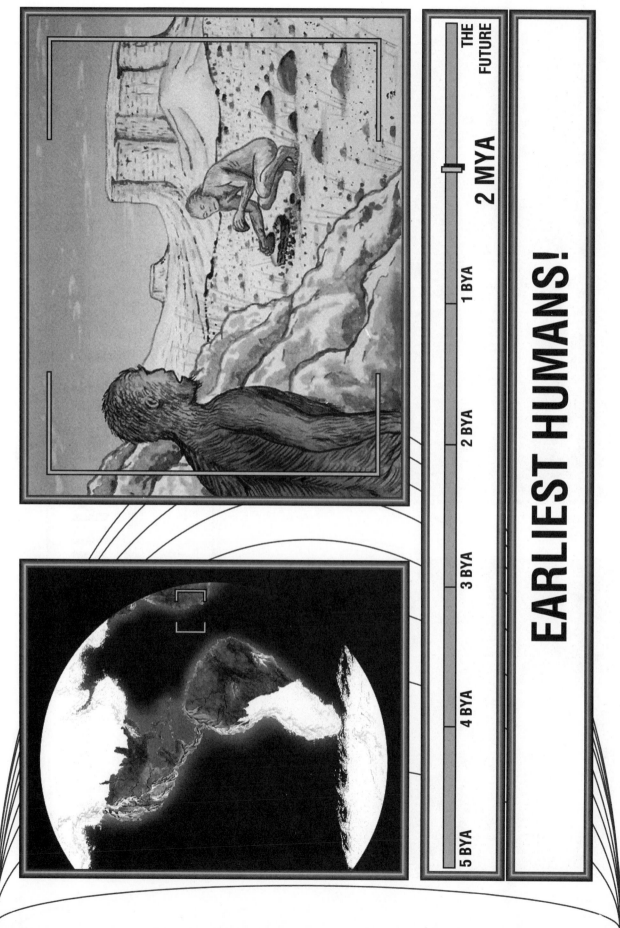

EARLIEST HUMANS!

5 BYA 4 BYA 3 BYA 2 BYA 1 BYA **2 MYA** THE FUTURE

From How Might Life Evolve on Other Worlds? © 1995. Teacher Ideas Press. (800) 237-6124.

5 BYA   4 BYA   3 BYA   2 BYA   1 BYA   PRESENT   THE FUTURE

NOW! CHILDREN OF THE EARTH HERE!

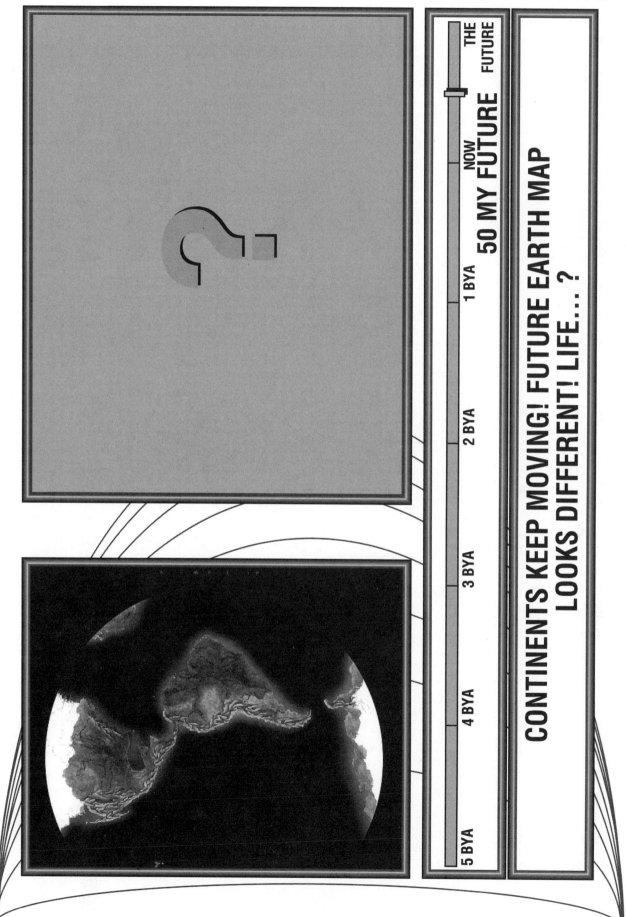

CONTINENTS KEEP MOVING! FUTURE EARTH MAP LOOKS DIFFERENT! LIFE... ?

5 BYA | 4 BYA | 3 BYA | 2 BYA | 1 BYA | NOW | THE FUTURE

50 MY FUTURE